Environmental Control Systems

An Independent Learning Module
from the
Instrument Society of America

24909021

ENVIRONMENTAL CONTROL SYSTEMS

By Randy D. Down

INSTRUMENT SOCIETY OF AMERICA

INSTRUMENT SOCIETY OF AMERICA
67 Alexander Drive
P.O. Box 12277
Research Triangle Park
North Carolina 27709

Library of Congress Cataloging in Publication Data
Down, Randy D.
Environmental control systems / by Randy D. Down.
p. cm. — (An Independent learning module from the Instrument Society of America)
Includes bibliographical references.
ISBN 1-55617-333-4
1. Factory and trade waste—Environmental aspects—Measurement—Programmed instruction. 2. Pollution control equipment—Programmed instruction. 3. Environmental monitoring—Programmed instruction. I. Instrument Society of America. II. Title. III. Series.
TD897.D68 1992 91-40221
628.5′0287—dc20 CIP

Editorial development and book design by Monarch International, Inc.

TABLE OF CONTENTS

SECTION THREE: SOIL

PREFACE

ISA's Independent Learning Modules

This book is an Indepedent Learning Module (ILM) as developed and published by the Instrument Society of America (ISA). The ILMs are the principal components of a major educational system designed primarily for independent self-study. This comprehensive learning system has been custom designed and created for ISA to more fully educate people in the basic theories and technologies associated with applied instrumentation and control.

The ILM System is divided into several distinct sets of Modules on closely related topics; such a set of individually related Modules is called a Series. The ILM System is composed of:

- the ISA Series of Modules on Control Principles and Techniques;
- the ISA Series of Modules on Fundamental Instrumentation;
- the ISA Series of Modules on Unit Process and Unit Operation Control;
- the ISA Series of Modules for Professional Development;
- the ISA Series of Modules for Specific Industries; and
- the ISA Series of Modules on Software-Assisted Topics.

The principal components of the Series are the individual ILMs (or Modules) such as this one. They are especially designed for independent self-study; no other text or references are required. The unique format, style, and teaching techniques employed in the ILMs make them a powerful addition to any library.

The published ILMs are as follows:

Fundamentals of Process Control Theory—Paul W. Murrill—1981

Controlling Multivariable Processes—F. G. Shinskey—1981

Microprocessors in Industrial Control—Robert J. Bibbero—1982

Measurement and Control of Liquid Level—Chun H. Cho—1982

Control Valve Selection and Sizing—Les Driskell—1983

Fundamentals of Flow Measurement—Joseph P. DeCarlo—1984

Intrinsic Safety—E. C. Magison—1984

pH Control—Gregory K. McMillan—1985

FORTRAN Programming—James M. Pruett—1986

Introduction to Telemetry—O. J. Strock—1987

Application Concepts in Process Control—Paul W. Murrill—1988

Controlling Centrifugal Compressors—Ralph L. Moore—1989

CIM in the Process Industries—John W. Bernard—1989

Continuous Control Techniques for Distributive Control Systems—Gregory K. McMillan—1989

Temperature Measurement in Industry—E. C. Magison—1990

Simulating Process Control Loops Using BASIC—F. G. Shinskey—1990

Tuning of Industrial Control Systems—Armando B. Corripio—1990

Computer Control Strategies—Albert A. Gunkler and John W. Bernard—1990

Fundamentals of Process Control Theory, Second Edition—Paul W. Murrill—1991

Environmental Control Systems—Randy D. Down—1992

Most of the original ILMs were envisioned to be the more traditional or fundamental subjects in instrumentation and process control. With the publications planned over the next few years, the ILM Series will become much more involved in emerging technologies.

ISA has increased its commitment to the ILM Series and has set for itself a goal of publishing four ILMs each year. Obviously, this growing Series is part of a foundation for any professional library in instrumentation and control. The individual practitioner will find them of value, of course, and they are a necessity in any institutional or corporate library.

There is obvious value in maintaining continuity within your personal set of ILMs; place a standing purchase order with ISA.

Paul W. Murrill
Consulting Editor, ILM Series

Comments about This Volume

Air, water, and soil are natural resources vital to our existence. As civilization on our planet has grown and industrialized, its impact on the environment has greatly increased.

Serious public and private concern over the long-term consequences of waste emissions on the natural environment has prompted the establishment of strict government legislations that closely regulate toxic emissions in order to greatly reduce, if not reverse, the impact.

This Independent Learning Module was developed to provide the reader with an understanding of instrumentation and control systems used to monitor and limit industrial and commercial pollutant emissions and their impact on the environment.

Dedication

I would like to dedicate this book to:

my parents, who always encouraged me to do my best;

my family, who have patiently supported me throughout the development of this book as well as my other career endeavors; and

the Instrument Society of America for the opportunity to create this text and for the enjoyment I have received being an active member of the Central New York Section and a splendid organization.

My Sincere Appreciation,

Randy D. Down, P.E.

Unit 1:
Introduction and Overview

UNIT 1

Introduction and Overview

Welcome to ISA's Independent Learning Module Environmental Control Systems. The first unit of this self-study program provides information on what is needed to take the course.

Learning Objectives — When you have completed this unit, you should:

A. Understand the general organization of the course.

B. Know the course objectives.

C. Know how to proceed through the course.

1-1. Course Coverage

This is a survey ILM on measurement and control systems as they apply to air and water emission sources that impact our natural environment. This course covers:

A. Basic concepts used to apply existing process control technology to the measurement and control of waste emissions.

B. Effective methods for measuring the impact of waste emissions on the surrounding air, soil, and water.

C. Portable, stationary, and remote (laboratory) instrumentation and their applications.

Technology related to the measurement and control of waste emission sources will continue to develop as government regulations become more stringent and more cost-effective and more reliable methods are developed. This text deals primarily with the available technology.

When applying techniques described within this text, it must be understood that Federal, State, and local government agencies should be consulted regarding permissible levels of toxic emissions and the regulations that apply to their measurement and control in your region of the country.

Appendix C identifies some of the governing agencies that should be consulted. Since policies and restrictions may vary greatly by region and do undergo frequent changes, it is impossible to accurately define them in this text.

1-2. Purpose

The purpose of this ILM is to present, in easily understood form, typical sources of plant emissions that impact the environment and methods to effectively monitor and control them. Applications identified in this text apply to both industrial and commercial situations.

Portable, stationary, and laboratory analytical instrumentation are all discussed because control projects often require some level of familiarity with all three.

Overall pollution control systems are discussed in some detail since it is very important to understand the process one is attempting to measure and control.

1-3. Audience and Prerequisites

This ILM is designed for those who wish to work on their own and want to gain a basic introductory understanding of environmental measurement and control systems.

The material will be useful to plant and facilities engineers and designers, environmental scientists, technicians, and consultants who are concerned with the environmental impact of new and existing processes. The course will also be helpful to students in technical schools, colleges, and universities who want to gain some insight into practical concepts of environmental monitoring and control.

No elaborate prerequisites are required to take this course, although some familiarity with basic instrumentation and control fundamentals would be useful. Other books in the ILM series do an excellent job of presenting these fundamentals.

1-4. Study Material

This textbook is the only study material required for this course. It is an independent, stand-alone textbook uniquely and specifically designed for self-study.

A list of suggested reading materials found in Appendix A will provide additional reference and study material.

1-5. Organization and Sequence

This ILM is divided into three sections. Each section is subdivided into separate study units. The next two units (2 and 3) are designed to teach the student about sources, measurement, and control of air emissions and indoor air quality. Units 4 and 5 address sources, measurement, and control of wastewater and groundwater contamination. The final units (6 and 7) cover soil analysis and remediation.

The primary method of instruction within this text is self-study. Basically, you will work on your own in taking this course; you select the pace at which you learn best in progressing through the course.

Each unit is designed in a consistent format with a set of specific learning objectives stated at the beginning of the unit. Note these objectives carefully. The material that follows the learning objectives will teach to these objectives. The individual units also contain exercises at the end to test your understanding of the material. The solutions to all student exercises are contained in Appendix E.

This ILM belongs to you; it is yours to keep as a learning tool and reference. We encourage you to make notes, taking advantage of the ample free space available on every page for this purpose.

1-6. Course Objectives

When you have completed this entire ILM, you should:

A. Be familiar with major sources of air, water, and soil contamination from industrial and commercial facilities and processes.

B. Understand the potential impact of these emissions on the environment and some of the regulations that govern them.

C. Be aware of the various portable and continuous measurement instruments available to monitor and analyze emissions.

D. Have an appreciation of control methods available to limit or eliminate emissions at their source.

E. Have an appreciation of methods available to recover chemical contaminants from process emissions for disposal or recycling.

F. Be familiar with some of the laboratory methods used to identify and measure contaminants.

1-7. Course Length

The basic premise of the ISA System of ILMs is that students learn best if they proceed at their own individual pace. As a result, there will be significant variation in the amount of time taken by individual students to complete the ILM. Previous experience and personal capabilities will do much to vary the time, but most students will complete this course within 20 hours.

Unit 2: Air Emission Sources

UNIT 2

Air Emission Sources

This unit identifies typical sources of pollutant air emissions from industrial and commercial facilities as well as methods to monitor and control them. While such sources as vehicle exhaust emissions are significant contributors to air pollution, they are not specifically addressed in this book. However, some of the techniques used to reduce pollutant emissions from vehicles, such as catalytic convertors, are also applied to industrial and commercial facilities and *are* described in this unit.

This text, in general, addresses emissions from what the U.S. Environmental Protection Agency (EPA) defines as ''stationary sources.''

In order to design an adequate measurement and control system for air emissions, a basic understanding of the government regulations and health concerns that drive the need for such a system should be understood. In addition, a basic understanding of the overall pollution control process is necessary in order to understand how instrumentation and controls related to this type of application must perform. In other words, it is difficult, if not impossible, to design appropriate measurement or control systems for an application without some understanding of the overall pollution control process.

Therefore, in addition to describing instrumentation for pollution monitoring and control systems, an overview of the overall control process is included.

Learning Objectives — When you have completed this unit, you should:

A. Understand what current government regulations apply to air emissions.

B. Be familiar with typical sources of air emissions.

C. Be aware of typical methods and equipment needed to monitor and record air emissions.

D. Understand methods of controlling air pollutant emission levels.

2-1. Federal and State Regulations

The U.S. Environmental Protection Agency (EPA) first adopted ambient air quality standards (pollutant concentration limits) under the Clean Air Act of 1970 in its initial approach to controlling air quality in order to protect human health and welfare.

Prevention of Significant Deterioration (PSD)

Congress, in directing the EPA to establish pollutant concentration limits, did not intend to allow air quality in geographical areas where it was already superior to established limits to degrade to levels specified under the Act. To ensure that this did not occur, the EPA developed further regulations for the prevention of significant air quality deterioration (PSD).

These PSD regulations became effective in December, 1974.

Amendments to the Clean Air Act that included PSD regulations were signed into law on August 7, 1977.

The following outline summary of PSD regulations provides some idea of the Federal limitations placed on pollutant emission sources. It is *not* intended to be a complete review and should not be used as a reference of Federal regulations. Current regulations may be obtained by directly contacting the EPA representative for your region and requesting this information. Refer to Appendix C for the address and telephone number of your region's representative.

PSD increments establish the maximum increase in pollutant concentration levels allowable above a defined baseline level. This baseline, referred to as the National Ambient Air Quality Standard (NAAQS), is shown in Table 2-1.

Averaging Period	Primary Standard	Secondary Standard
Sulfur dioxide		
Annual	0.03 ppm	none
24-hour	0.14 ppm	none
3-hour	none	0.5 ppm
Particulate matter		
Annual	75 $\mu g/m^3$	60 $\mu g/m^3$
24-hour	260 $\mu g/m^3$	150 $\mu g/m^3$
Carbon monoxide		
8-hour	9.00 ppm	9.00 ppm
1-hour	25.00 ppm	25.00 ppm
Ozone		
1-hour	0.12 ppm	0.12 ppm
Nitrogen oxide		
Annual	0.055 ppm	0.05 ppm
Lead		
Calendar qtr.	—	
Hydrocarbons		
3-hour	0.24 ppm	0.24 ppm

Table 2-1. National Ambient Air Quality Standards (Source: U.S. Environmental Protection Agency)

The only pollutants Congress initially regulated under this incremental approach were sulfur dioxide and particulate matter. At the same time, the EPA was directed by Congress to determine whether the need existed to establish a similiar approach for regulating other air pollutants.

In June 1978, an entirely new set of PSD regulations were adopted by the EPA. The new regulations identified new increments mandated by Congress, declaring what types of emissions sources were now subject to PSD regulations, and also covering other issues, such as Best Available Control Technology (BACT) and ambient monitoring requirements as directed under the Clean Air Act.

Several provisions in the 1978 PSD regulations were challenged by a combination of environmental and industrial groups. A ruling by the U.S. District Court of Appeals resulted in revisions to these regulations in August of 1980.

Stationary Sources

Under PSD regulations, a ''stationary source'' is defined as ''all pollutant-emitting activities, which occur in the same industrial grouping, are contained at one location (one or more contiguous or adjacent properties), and are owned or under the control of the same person or persons.''

Major new sources require a PSD permit; construction cannot begin without this permit. Determining whether a source is considered ''major'' is based on its maximum potential emissions with its *control systems operating.*

Major Sources

A list of specific sources with potential emissions of 100 tons per year or more of a regulated pollutant are identified within PSD regulations as ''major'' sources. These 28 source types include kraft pulp mills, iron and steel mills, petroleum refineries, fuel conversion, and sulfur recovery plants. Any other sources with potential emissions of 250 tons per year or more of a regulated pollutant are also considered major sources.

A PSD permit is also required for any *existing* major source that creates a physical change in its method of operation that results in a ''significant'' net increase in air emission levels (see Table 2-2).

Significant emissions or emissions increase are identified under PSD regulations as:

Pollutant	Emissions Rate (Tons per Year)
Carbon monoxide	100
Nitrogen oxides	40
Sulfur dioxide	40
Particulate	25
Ozone	40 (VOCs)
Lead	0.6
Asbestos	0.007
Beryllium	0.0004
Mercury	0.1
Vinyl chloride	1
Fluorides	3
Sulfuric acid (mist)	7
Hydrogen sulfide	10
Total reduced sulfur	10
Reduced sulfur compounds	10

Table 2-2. PSD Significant Emissions (Source: U.S. Environmental Protection Agency)

A PSD review is required if there is a significant increase in any pollutant, not just pollutants from a major source.

Modifications at an existing ''minor'' source do not require a PSD permit.

The cumulative effects of fugitive emissions are also included in determining whether a source is major if they are emitted from one of the specific major sources identified earlier under PSD regulations (the 28 source types).

2-2. Stack and Exhaust Emissions

Any gases emitted into the atmosphere that are considered toxic fall under regulations established under the Clean Air Act, as well as State and local government regulations. This section addresses emissions that enter the atmosphere and are not considered ''fugitive.'' This generally includes toxic gases exhausted from a facility or process through chimneys, stacks, or exhaust fans. These toxic gases are typically created as by-products of some type of combustion or chemical process.

Combustion By-products

Gaseous pollutants found in combustion exhaust emissions tend to be difficult to remove. Particularly harmful environmental pollution can result if high levels of gaseous oxides of nitrogen and sulfur are emitted from fossil fuel-burning plants.

Sulfur oxides result from the presence of sulfur in fossil fuels. Relatively low levels of sulfur are also found in natural gas (see Table 2-3). Ninety-eight percent of the sulfur oxides created by combustion occur in the form of sulfur dioxide or SO_2.

Fuel	Average Sulfur Content (Percent by Weight)
High sulfur coal	16%
Low sulfur coal	12%
No. 6 fuel oil	4%
No. 4 fuel oil	3%
No. 2 fuel oil	2%
Natural gas	>1%

Table 2-3. Average Sulfur Content (Source: *The Control of Boilers*, Dukelow, Sam G., ISA, 1986.)

Acid Rain

High concentrations of sulfur dioxide allowed to enter the atmosphere have been blamed for "acid rain." Acid rain develops when moisture in the atmosphere combines with sulfur dioxide to form sulfuric acid. Acidic rainfall gradually changes the pH level of surface water and soil. The full impact of this change on the natural environment is still a subject of some debate. An explanation of pH can be found in Unit 4.

2-3. Fugitive Emissions

Sources

Fugitive emissions are pollutant emissions from sources that occur as a result of leaks around pipe fittings, valve stems, air vents, and other openings in a containment system. These allow small quantities of vapors, liquid, or dust to leak and disperse into the surrounding environment. It would be nearly impossible to prevent all fugitive emissions. However, the cumulative effect of these numerous small leaks can be quite significant in a large process facility and should be addressed.

CFCs

A good example of one type of fugitive emission that has received much public attention is chlorofluorocarbon, or CFC. CFCs are compounds composed of carbon, chlorine, and fluorine. Examples of CFC uses include refrigeration (in the form of Freon™), air conditioning, solvents, plastic foam (for insulation or packing materials), sterilants, and aerosol propellants (banned for this purpose in the U.S. in 1978). CFCs are odorless, colorless, nonflammable, noncombustible, synthetic chemicals with a low toxicity level, making them attractive substances for widespread applications.

Some forms of this widely used synthetic chemical compound have been linked to depletion of the earth's ozone layer. The ozone layer is credited with absorbing much of the ultraviolet radiation from the sun, thus protecting plant and animal life on earth. Life expectancy of CFC compounds in the earth's atmosphere is estimated to be 75 to 185 years. Fugitive emissions from refrigeration and air conditioning systems contribute a significant amount of the CFCs released into the atmosphere. It is believed that a worldwide ban of CFC production may be enacted by the year 2000.

Low-Leakage Valve Technology

Small process gas or liquid leaks around valve stem packings or seals can be a significant source of fugitive emissions when one considers the large number of manual and automatic valves found at a typical industrial facility. Fugitive emissions of toxic substances potentially risk personnel safety, contribute to pollution of the environment, and reduce productivity and efficiency of the plant.

Leakage rates from valves, like other sources of fugitive emissions, are largely a function of the process variables to which the valve is applied, such as temperature, pressure, and material compatability.

Most contemporary valve designs have relatively low leakage rates. The EPA's definition of a "leak" is 10,000 parts per million. In many cases, this is an acceptable level in terms of personnel safety and efficiency. However, in applications where the leaking fluid is highly toxic, such as cadmium or cyanide, this level may be totally unacceptable. In such a case, a "zero leakage" valve must be used.

When selecting valves for any process application involving toxic substances and pollutants, the design of valve stems, seals, and packings must be carefully considered (see Fig. 2-1).

The following low-leakage valve designs are considered effective in nearly eliminating fugitive emissions at valve stem seals.

Bellows-Sealed Valves

Bellows-sealed valves (see Fig. 2-2) work well in a limited size range and materials of construction. They are most frequently used on valve sizes of two inches or less. The practical physical limitation of this type of seal is on eight-inch valves. Bellows designs typically have a limited service life expectancy, which should be given consideration during selection. Frequency of valve stem movement is determined by the type of application, the control method, and tuning parameters that govern valve

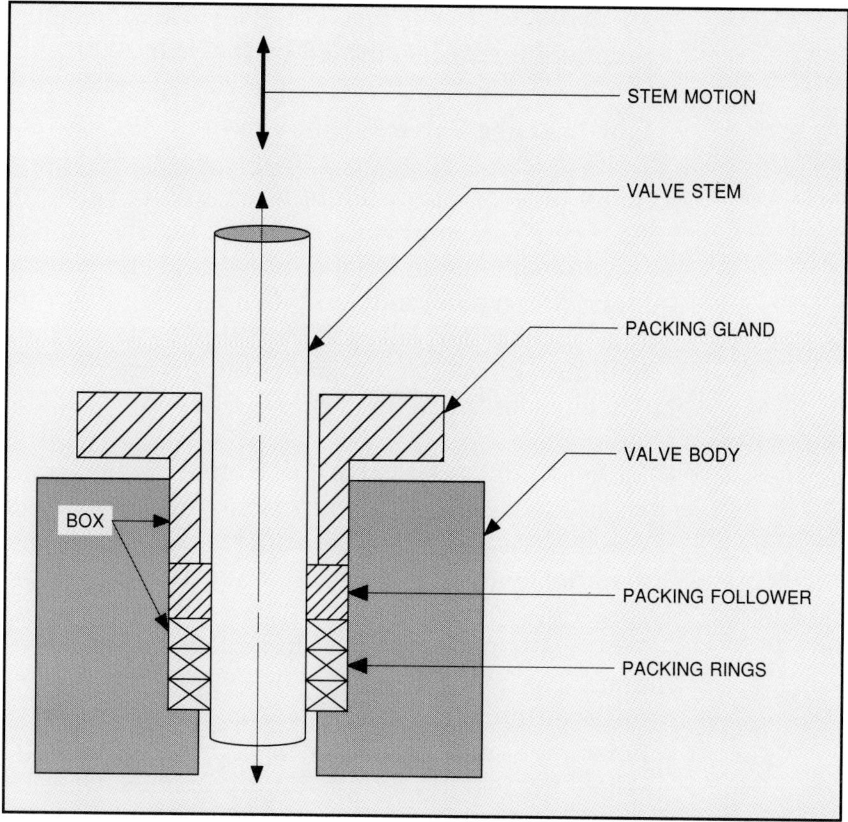

Fig. 2-1. Standard Valve Packing

modulation. The most common application for this type of valve is on steam services.

Diaphragm Valves

A diaphragm valve basically consists of a valve stem connected to a diaphragm that flexes with stem motion (see Fig. 2-3). Deflection of the diaphragm varies the flow area of gas or liquid passing through the valve body. This design concept provides a natural hermetic seal, preventing leakage around the valve stem during operation since the stem is isolated completely from the process fluid.

The diaphragm is usually constructed of rubber, which limits its service capability to low-temperature applications. This type of valve is also available with thin metal diaphragms.

Inherently, metal fatigue limits the stroke length and life expectancy of these valves. However, in applications involving

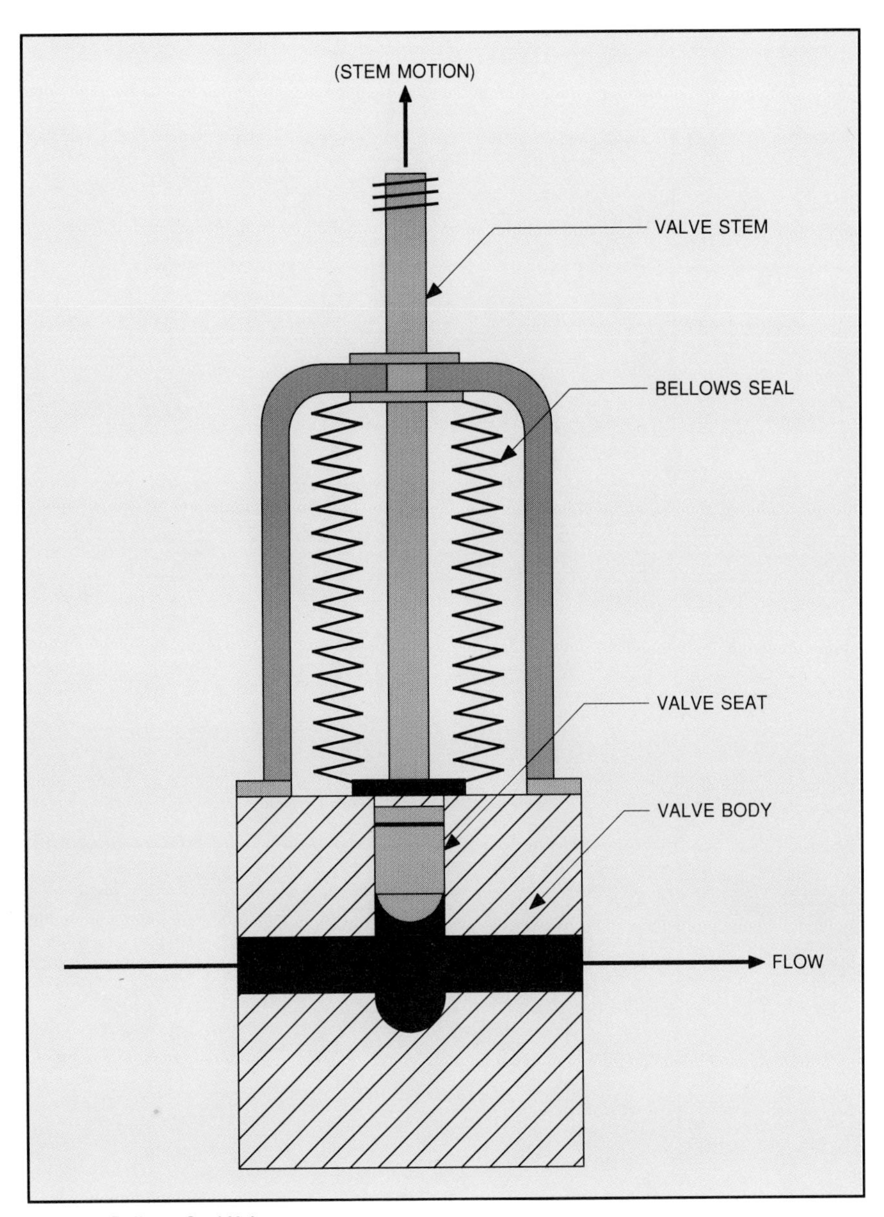

Fig. 2-2. Bellows Seal Valve

high process fluid temperatures and low frequency of valve modulation and requiring zero stem leakage, this type of valve construction may be ideally suited.

Pinch Valves

Another version similiar to the diaphragm design is the "pinch" valve. This design also provides a hermetic seal

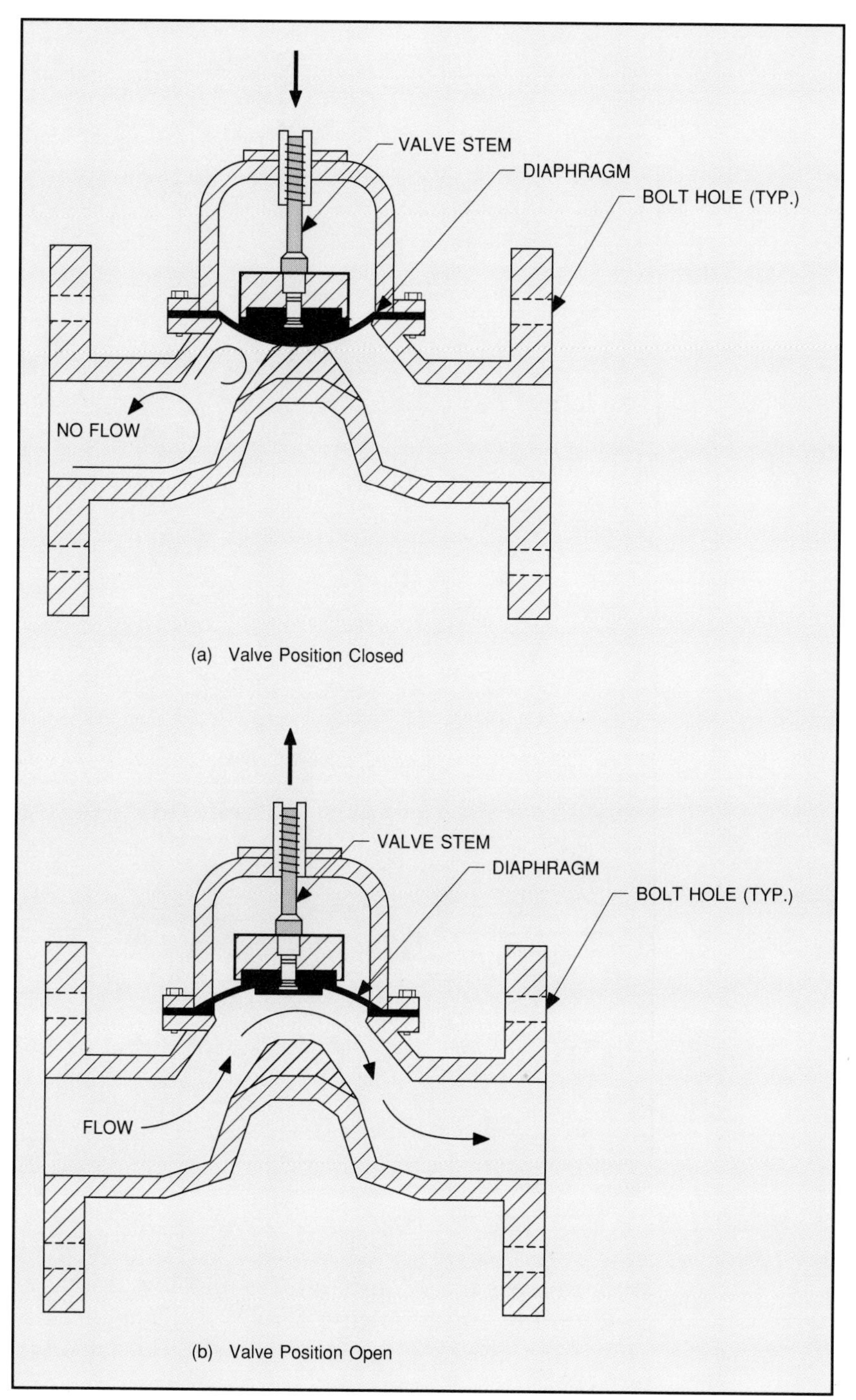

Fig. 2-3. Diaphragm Valve

between the controlled fluid and valve stem. A sleeve designed of flexible material is pinched by air or mechanical means to restrict the flow of fluid. These valves are particularly effective in applications involving slurries or sludge because of their ability to seal tightly in the presence of high viscosity materials and solids. Sleeve materials are available that can withstand temperatures up to 400°F.

Live Loading

The term "live loading" generally refers to the use of metal springs mounted on top of the packing gland to maintain a dynamic downward force on the valve packing at all times (see Fig. 2-4).

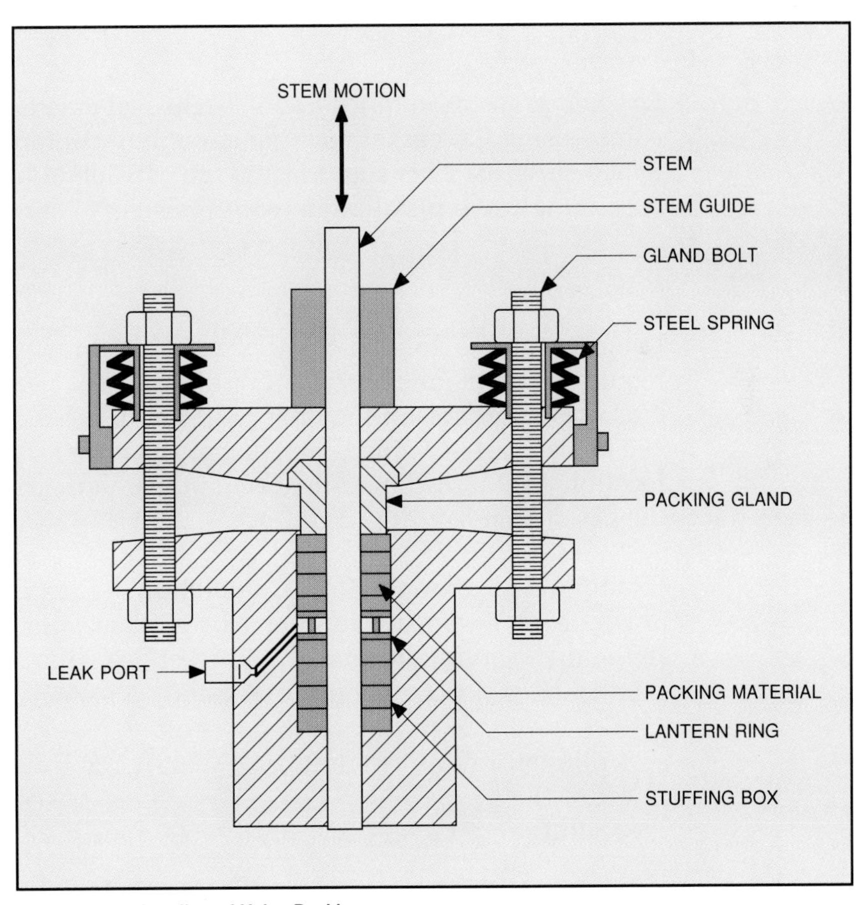

Fig. 2-4. Live Loading of Valve Packings

Dynamic characteristics of the springs allow the packing to compensate for variations in stem diameter that occur due to thermal expansion and stem wear. Live loading can be applied to most valve designs and is a cost-effective way of minimizing stem leakage.

Several valve manufacturers have developed and are refining the use of multiple-ring, live-loaded packings for low-leakage applications.

Pipe Fittings

Another potential source of fugitive emissions exists at pipe fitting connections. Proper gasketing of flanged connections provides a reliable, leakproof seal. Examples of three commonly used gasketing designs are shown in Fig. 2-5. Leakage occurs when insufficient or excess pressure is applied to the gasket material.

Leakage prevention relies upon a tight seal created by compressing the gasket material enough to conform to the flange surfaces. This should be given consideration when specifying valve installation requirements.

2-4. Monitoring Methods

Determining the volume of gases emitted into the atmosphere requires accurate measurement of the gas flow rate. Since the cross-sectional area of the duct or breaching ahead of the stack can be easily determined, measuring gas velocity at the same location will allow determination of the volumetric flow rate of the gas stream.

Example 2-1. Sample Flow Calculation. Determine the rate at which gas is exhausted from a stack if measured gas velocity in the duct serving the stack is 900 feet per minute (fpm) and the inside duct dimensions are 4 feet by 8 feet.

$$\text{Volumetric flow rate (cfm)} = \text{Area (ft}^2) \times \text{Velocity (fpm)}$$

$$\text{Solution: cfm} = (4 \text{ ft} \times 8 \text{ ft}) \times 900 \text{ fpm} = 28{,}800 \text{ cfm.}$$

A number of instruments are available to measure gas velocity, including Pitot tubes, anemometers, airfoils, venturis, and

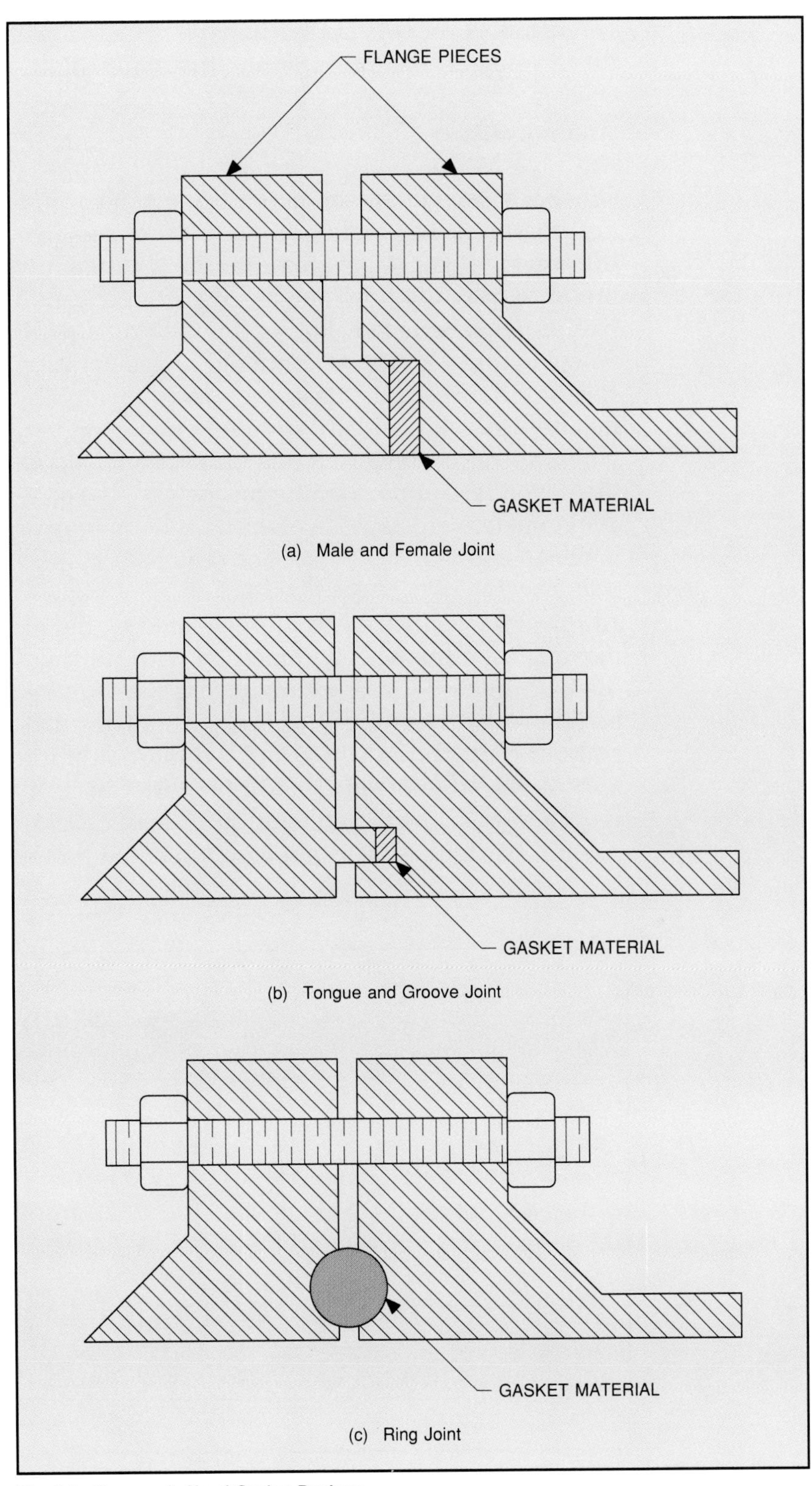

Fig. 2-5. Commonly Used Gasket Designs

variations of these basic methods. Each has its advantages and disadvantages, depending upon the application.

Anemometers

One basic type of anemometer is a turbine or propeller-driven unit that uses a propellor or a fan blade (see Fig. 2-6). When this anemometer is inserted into the air stream, velocity pressure from the flowing gas forces the wheel to spin, much like pinwheels that children play with on a windy day. The wheel spins at a rate proportional to gas velocity. Rotational speed of the wheel is measured electronically (by a magnetically or optically coupled circuit) on newer units. The wheel speed is converted to a proportional signal that is displayed in feet per minute (or meters per second). Turbine anemometers are capable of accurately measuring flows between 200 and 4,000 feet per minute.

At lower gas velocities, friction created by the mechanical movement of the wheel, combined with starting torque, cause the instrument to lose accuracy. The measured gas stream must be relatively dry and free of fine particulates that can collect and build up on the wheel and the interior of the wheel or turbine housing, creating additional friction and system inaccuracy. While suitable for quick spot checks of gases that contain particulate, it is not recommended that this instrument

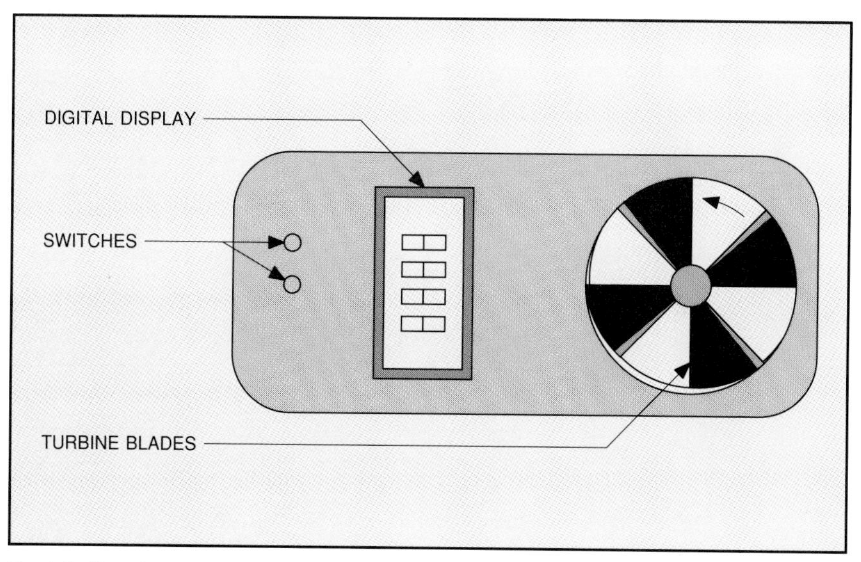

Fig. 2-6. Vane Anemometer

be used for continuous measurement because of the high frequency of maintenance required for proper operation.

The second type of anemometer, the hot-wire or hot-film type, determines air velocity by measuring the heat dispersion rate of the gas stream (see Fig. 2-7). The higher the velocity of gases passing over the hot wire or film, the more heat carried away by the gas. Since heat dispersion rate also varies slightly with gas temperature (gas density varies with temperature), gas temperature is simultaneously measured by an integral thermistor or thermocouple mounted in the velocity probe of the anemometer. Temperature variations are compensated for electronically within the unit.

Hot-wire or film anemometers are accurate within ±3% of full scale on low temperature gas streams for velocities ranging from 100 to 6,000 feet per minute. On applications above a gas temperature of approximately 175°F, alternative methods of measurement are more effective.

Another advantage of hot-wire units versus the turbine design is the probe diameter. While turbine anemometers are available with much smaller wheel diameters than that shown in Fig. 2-6, a probe on the hot-wire version can be as small as 5/8th inch in diameter. Therefore, a smaller test port is required in the wall of the duct.

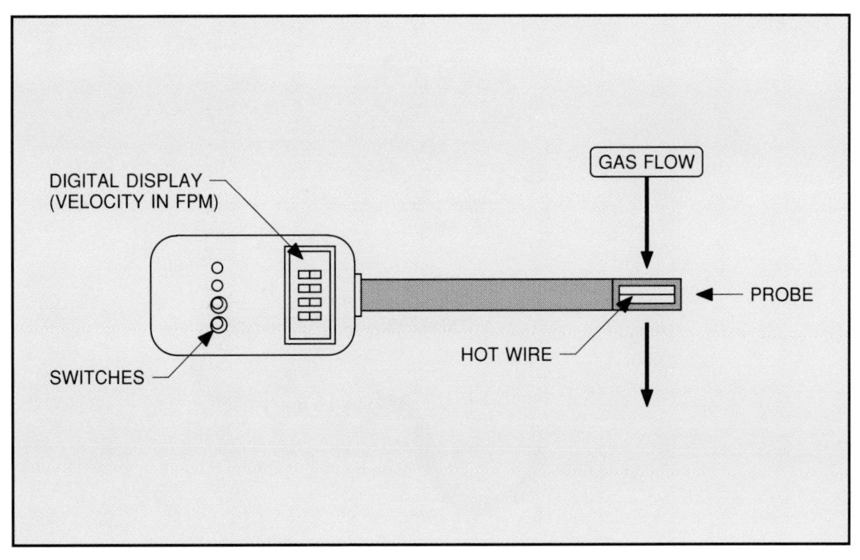

Fig. 2-7. Hot-wire Anemometer

Pitot Tube

Pitot tubes are commonly used throughout industry for the measurement of gas flow. They are relatively inexpensive, simple to use, and reliable when used in low particulate, dry, clean gas applications. They also create very little pressure drop in the gas stream.

The basic instrument consists of a double-wall perforated tube connected to some type of low differential pressure gage, transmitter, or manometer. The probe is designed with a curved tip that allows the open end of the inner tube to be turned directly into the path of the gas stream to measure static pressure. The second (outer) tube has a series of small perforations that allow total static (stagnation) pressure in the duct to be measured. The differential pressure (or pressure difference) between measured static pressure and total pressure represents velocity pressure as displayed on the pressure indicator (see Fig. 2-8).

Gas velocity is proportional to the square root of the velocity pressure and inversely proportional to gas density.

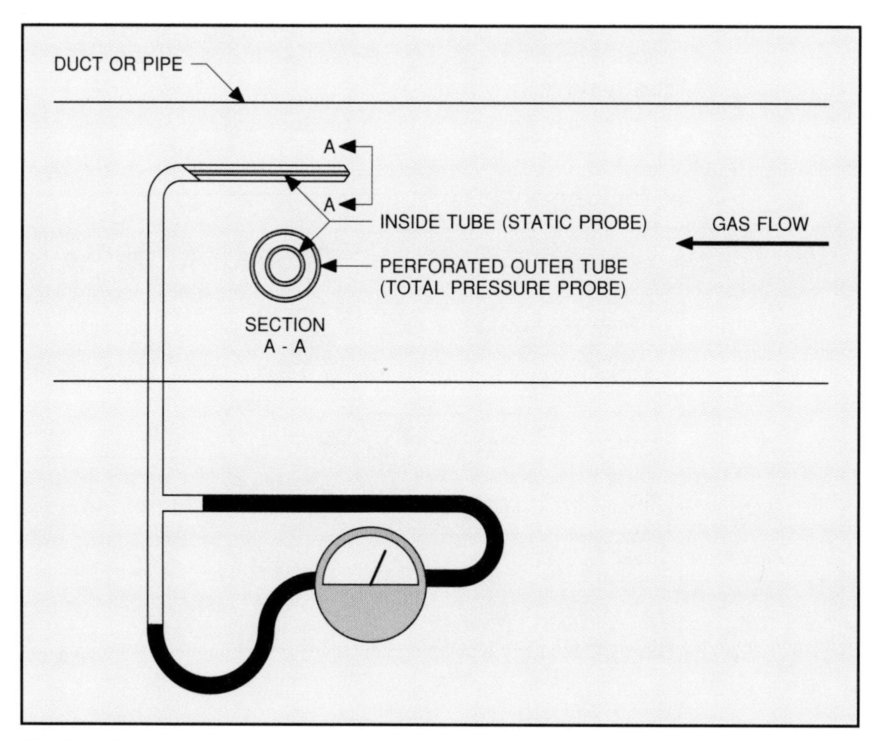

Fig. 2-8. Pitot Tube

Pitot tubes have no mechanical movement, but are susceptible to fine particulates and viscous contaminants in the gas stream. Oil vapor or particulate in the gas stream can plug the very small pressure-sensing ports in the outer tube, rendering the instrument ineffective. By equipping the Pitot tube with an instrument air purge connection, to periodically blow air back out through the sensing ports, this problem can be substantially reduced.

All velocity measurement methods are somewhat susceptible to fouling in dirty wet-gas streams, since the wet particulate will tend to cling and collect on whatever object it comes into contact with.

Pitot tube measurement accuracy depends greatly upon the accuracy of the pressure measurement instrument to which it is attached. Micromanometers with an accuracy of 0.001 inch of water column are capable of accurately measuring air velocities down to 900 feet per minute using a Pitot tube.

It is often difficult to find a long, straight section of ductwork that will provide a nonturbulent (laminar) flow pattern for single-point velocity measurement. Therefore, in order to accurately determine the average velocity in a duct, traverse measurements of velocity must be obtained. Traverse readings are a series of measurements taken horizontally and vertically across the cross-sectional area of the duct (see Fig. 2-9). Measurements are then averaged together to determine the average velocity.

Permanent Pitot tube installations typically consist of a flow cross or a single bar that continuously averages the velocities across the duct by simultaneous measurement of the pressure points.

Differential Pressure Devices

The remaining flow measurement elements create a differential pressure (pressure drop) in the gas stream by placing an object in the stream and measuring the pressure difference on either side of the object (see Fig. 2-10). Flow velocity varies with the square of the pressure differential.

As in liquid flow measurement, orifice plates are used in gas streams for flow measurement. However, the relatively large

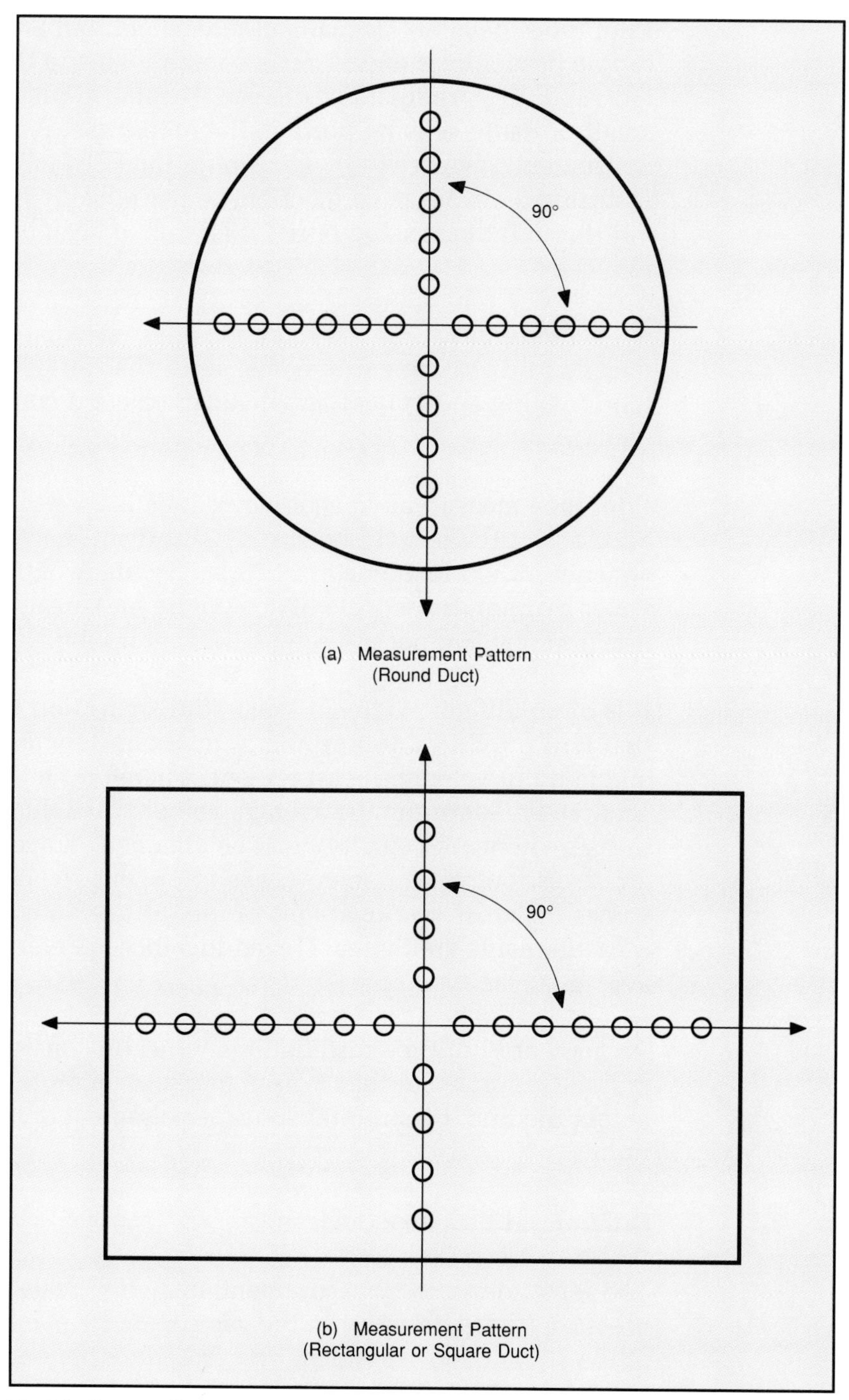

Fig. 2-9. Traverse Measurement Diagram

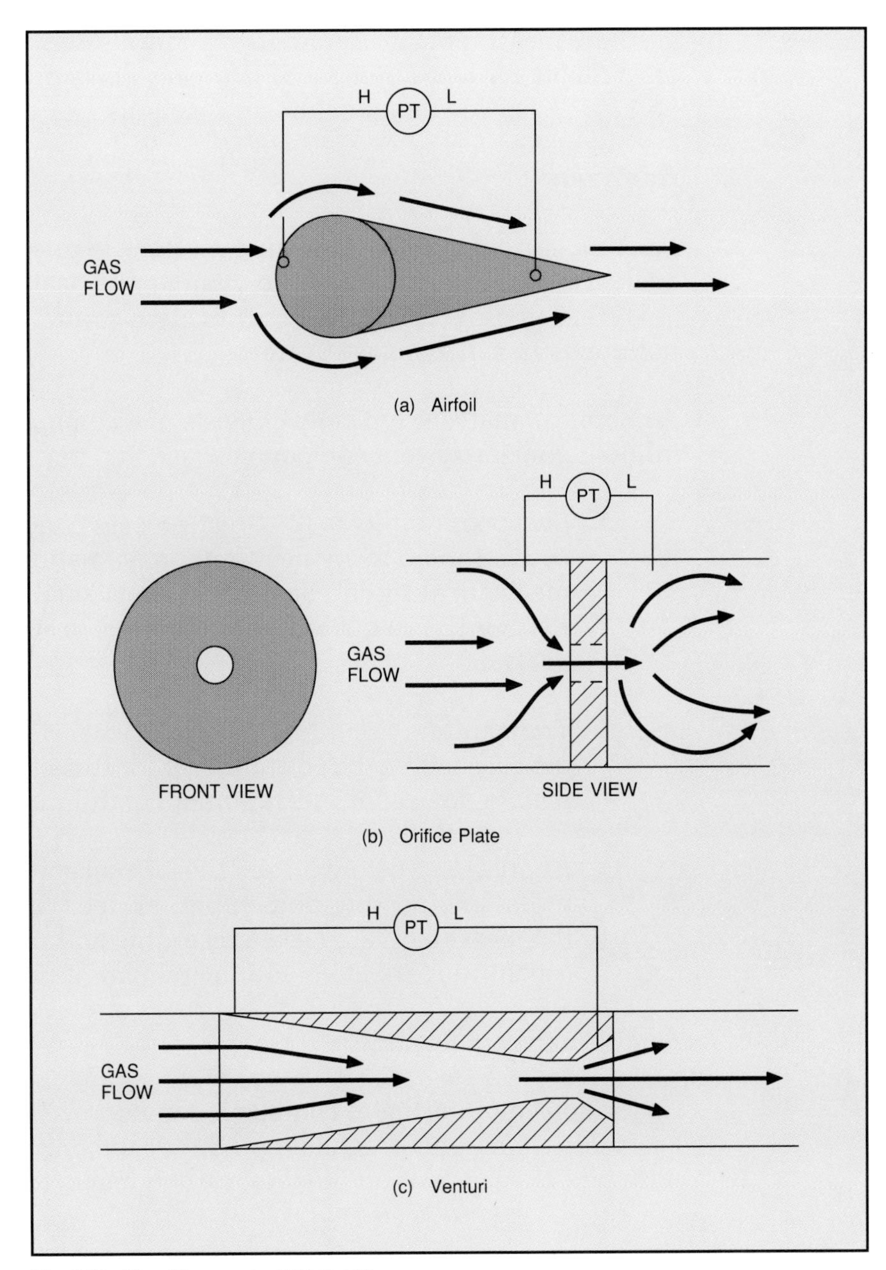

Fig. 2-10. Flow Measurement Elements

pressure drop created by the plate can result in a need for increased fan horsepower in order to offset the pressure loss. Over time, this could result in significant energy costs. Orifice plates also create a place for particulates to accumulate.

As an alternative, devices that create a more modest pressure drop and are somewhat self-cleaning have been designed for gas flow measurement. Aerodynamic airfoils and venturi tubes are suitable for measurement of large gas flows in both high (>175°F) and low temperature applications.

Analyzers

Meeting current and pending air emissions regulations requires analysis beyond temperature, flow rate, and opacity. Chemical analyzers are now being used to directly measure levels of toxic pollutants in exhaust gas streams.

A chemical analyzer typically consists of a combination of the following major system components (see Fig. 2-11):

A. *Sensors/Samplers.* To detect process conditions and convert them to an electronic, pneumatic, or mechanical signal. Samplers collect small representative quantities of the process stream for analysis.

B. *Transmitters/Conditioners.* To characterize (linearize) the sensor signal and amplify it for transmission to a remote device for indication or control purposes.

C. *Analyzers.* To convert the sensor/transmitter signal to useful engineering units (ppm, pH, etc.) and display the measured value. Data recording and process control capabilities are often an integral part of the analyzer.

Stack Analyzers for CEM

Continuous monitoring of combustion emissions using stack analyzers is not a new technology. Sulfur dioxide, oxides of nitrogen, oxygen, carbon dioxide, carbon monoxide, and

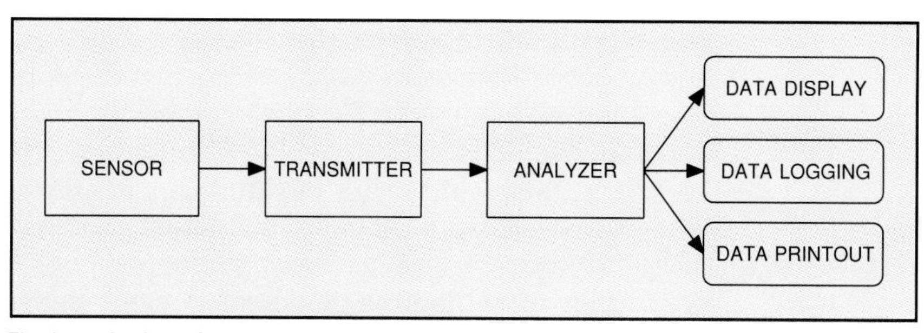

Fig. 2-11. Analyzer Components Block Diagram

opacity analyzers have been used successfuly for many years on utility boilers.

Monitoring of sulfur dioxide (SO_2), as an example, has aided in the blending of fossil fuels in order to minimize sulfuric acid created when hot sulfur-containing exhaust gases come into contact with cool, moist ambient air and condense (at approximately 260°F). The following equation represents the chemical reaction between water and sulfur oxides that creates sulfuric acid:

$$SO_3 + H_2O = H_2SO_4 \text{ (sulfuric acid)} \qquad (2\text{-}1)$$

Acid effluent contributes not only to acid rain but also to rapid corrosion damage of metallic flue ductwork and stack if the gas temperature drops below its dewpoint (condenses) before it is emitted to atmosphere.

Opacity Monitors

Opacity measurement tells the plant operator how dense smoke or particulate in the exhaust gases is. Huge billowing clouds of black smoke emitted from a stack are much more disconcerting to the general public because they can see it. Opacity measurement also assists the operator in determining how efficiently the boiler is being operated.

An opacity monitor projects a beam of visible light across the flue duct to either a mirror, which reflects the light beam back into the receiver unit, or directly to a receiver unit on the opposite side of the duct (see Fig. 2-12). The more dense the smoke or particulate concentration in the gas stream, the more light is dispersed and, therefore, does not reach the receiver. Measurement of this condition is scaled in ''percent opacity.''

In combustion control applications, an opacity monitor can be used as a high limit device that limits combustion when the density of particulates created by the combustion process approaches a defined limit. It does not work particularly well for combustion trim control or direct combustion control.

Reliability has been a problem in the past, particularly on coal-fired or incinerator applications that generate high volumes of fly ash. The problem centers around keeping the instrument's optical system clean enough to provide reliable measurement.

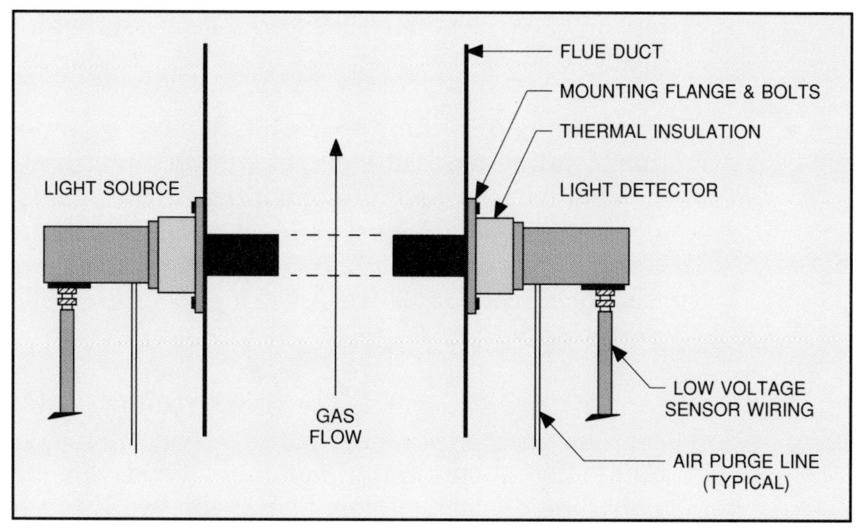

Fig. 2-12. Opacity Monitor

Particulate would rapidly build up on the lenses, blocking the light path and making the unit inaccurate until it was removed and cleaned. Quite often these monitors are located in very inaccessible locations (high up on the stack or breeching).

Newer units are equipped with blower assemblies that create a curtain of air in front of the glass lenses at the light source and detector to prevent fouling (see Fig. 2-13). With fuels such as natural gas and fuel oils, the fly ash concentrations are much lower; therefore, fouling is not a great concern.

Newer opacity monitors with a method to automatically correct for gradual fouling of the lenses are available. The system shown in the diagram of Fig. 2-14 can be equipped with an external fiber optic cable that acts as a reference for automatic calibration.

CEM for Combustion Efficiency Control

Oxygen, nitrogen, carbon dioxide, and carbon monoxide analyses are typically used independently, or in combination, to determine combustion efficiency (percentage of fuel actually burned) and, as mentioned earlier, provide a ‘‘trim’’ (fine tuning) adjustment to standard air/fuel ratio combustion control schemes. A typical O_2/CO combustion trim control arrangement

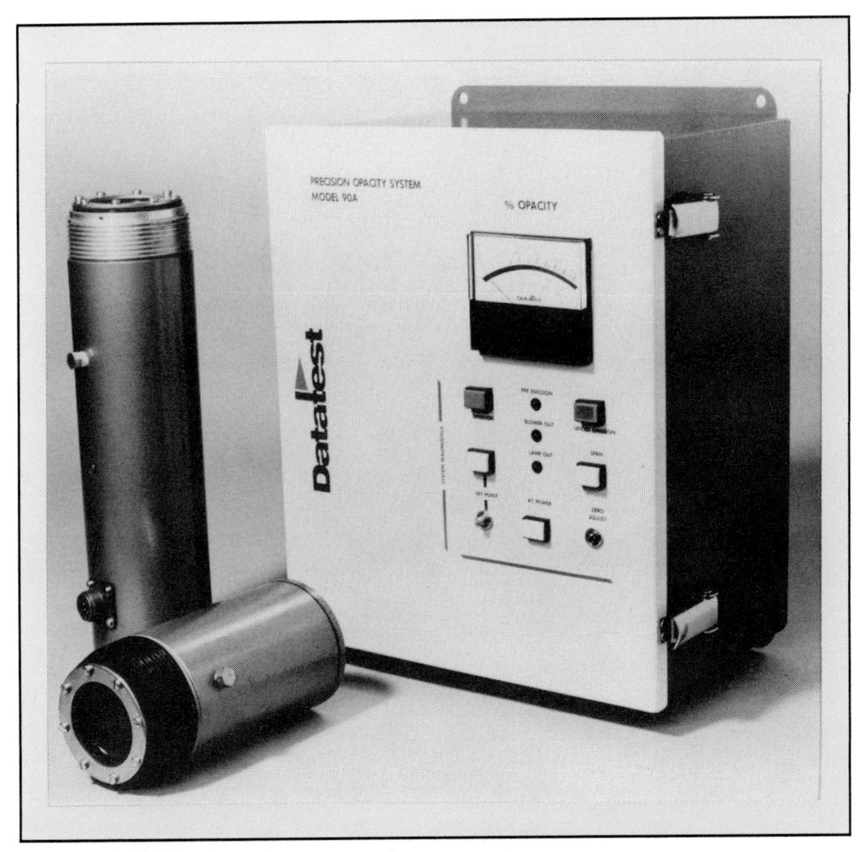

Fig. 2-13. Opacity Monitoring System (Courtesy of Datatest)

is represented in the SAMA diagram in Fig. 2-15. SAMA (Scientific Apparatus Maker's Association) diagrams are commonly used in the power industry to represent a control process in functional block diagram format. ISA has adopted some of the SAMA symbology into its standards.

On large field-erected boilers, such as those used by utilities, one to two percent improvements in combustion efficiency can save enough money through reduced fuel consumption to pay for the analyzers in a matter of months.

By industry standards, any return on investment (ROI) of less than three years is considered a worthwhile investment. This type of efficiency improvement can be accomplished through tighter burner control using combustion flue gas analyses (see Fig. 2-16).

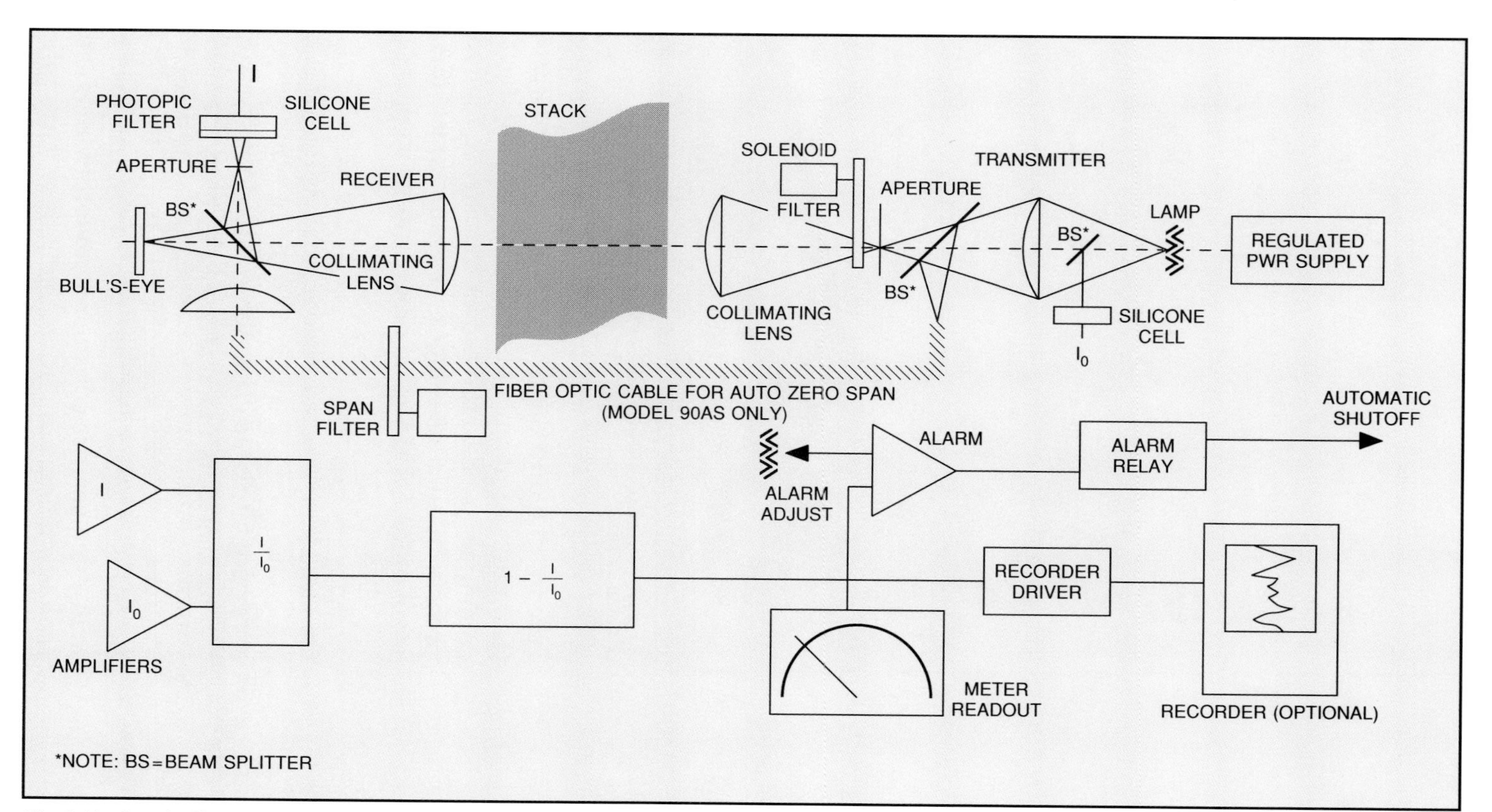

Fig. 2-14. Opacity Monitor with Fiber Optic Cable for Auto Zero and Span (Courtesy Datatest)

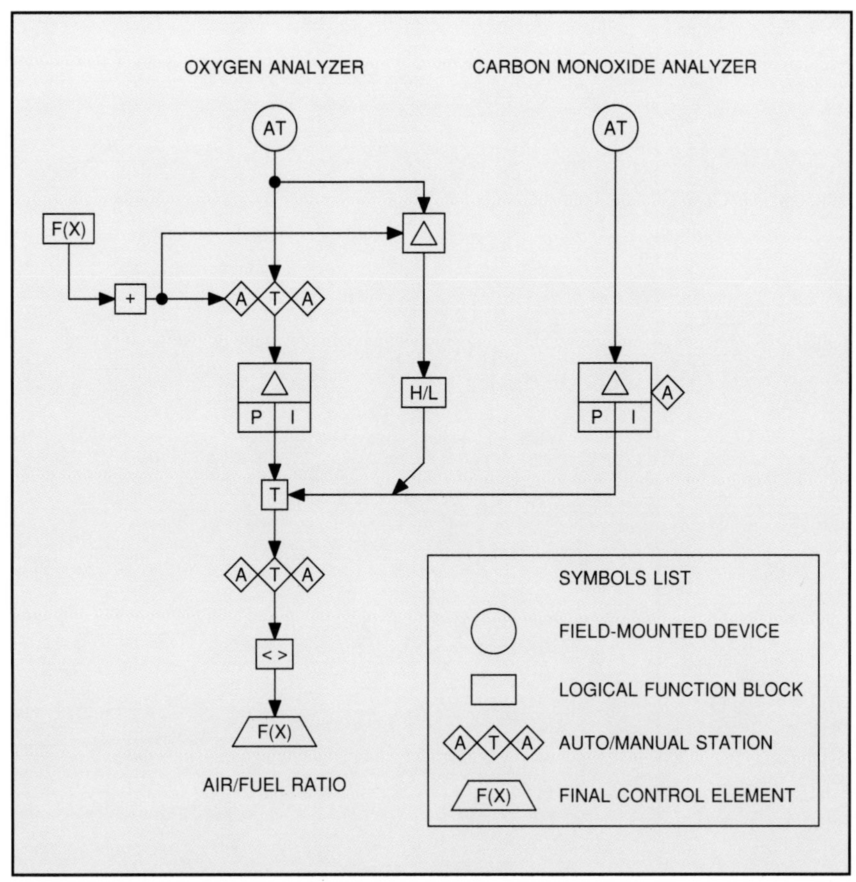

Fig. 2-15. SAMA Combustion Control Diagram (with CO/O_2 Trim)

On smaller field-erected or package boilers that consume less fuel, the potential savings will be somewhat less and should be determined before money is invested in analysis equipment for efficiency improvements.

When required by law to monitor and control pollutant emission levels, there is no choice but to invest in the equipment and comply with the regulations. In doing so, keep in mind that the accuracy, repeatability, and reliability of the analysis equipment must meet the standards established by the applicable regulatory agencies in your area. When in doubt, contact your regional representatives and ask before you invest a large sum of money for equipment that may not be necessary or is inadequate.

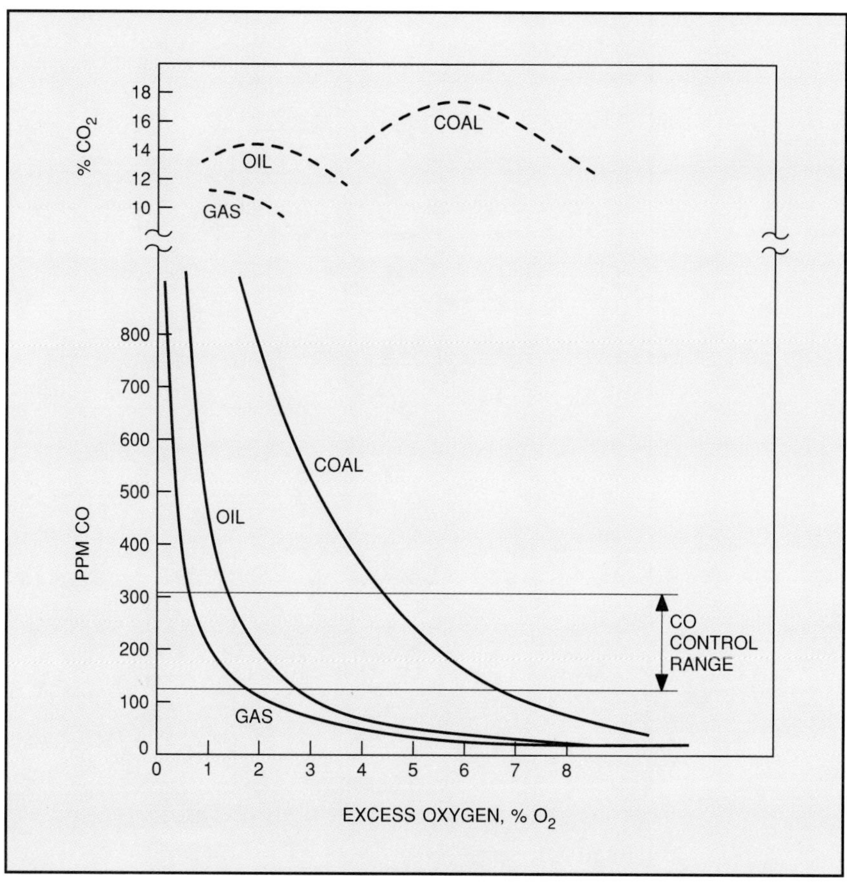

Fig. 2-16. Excess O_2 and CO or CO_2 in Fluegas of a Boiler That Was Operated at a Constant Load

In Situ versus Extraction Sampling

Continuous emissions monitoring (CEM) is accomplished using either in situ or extractive sampling techniques.

There has been significant debate by equipment manufacturers over the performance of the in situ-type sampling system versus that of the extractive system. A number of manufacturers recognize that both methods have their advantages, depending on the application, and now compromise by offering either version.

In situ units sense constituents in gas streams directly by insertion of a gas sensor directly into the stream (see Fig. 2-17).

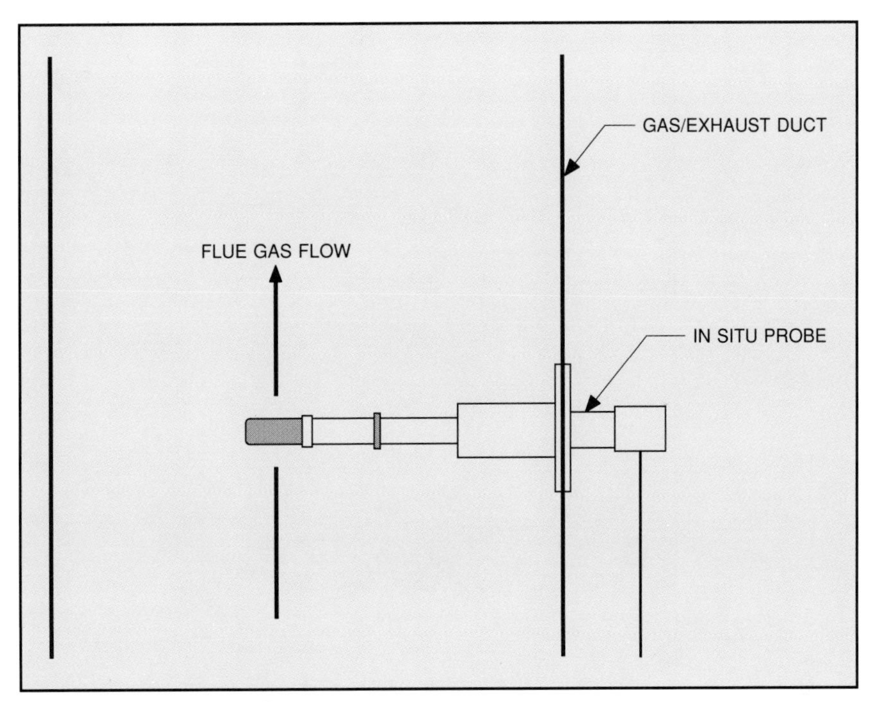

Fig. 2-17. In Situ Sampling Probe

In situ measurement offers the following advantages and disadvantages:

Advantages

A. Direct measurement under actual conditions.

B. Generally less expensive than extraction.

C. Less complex method of sampling than extraction.

Disadvantages

A. Often less accessible for maintenance.

B. Exposure to harsh operating conditions.

Dry versus Wet Sampling

Extraction systems sense constituents in gas streams by either a "dry" or a "wet" sampling process. In the dry process a

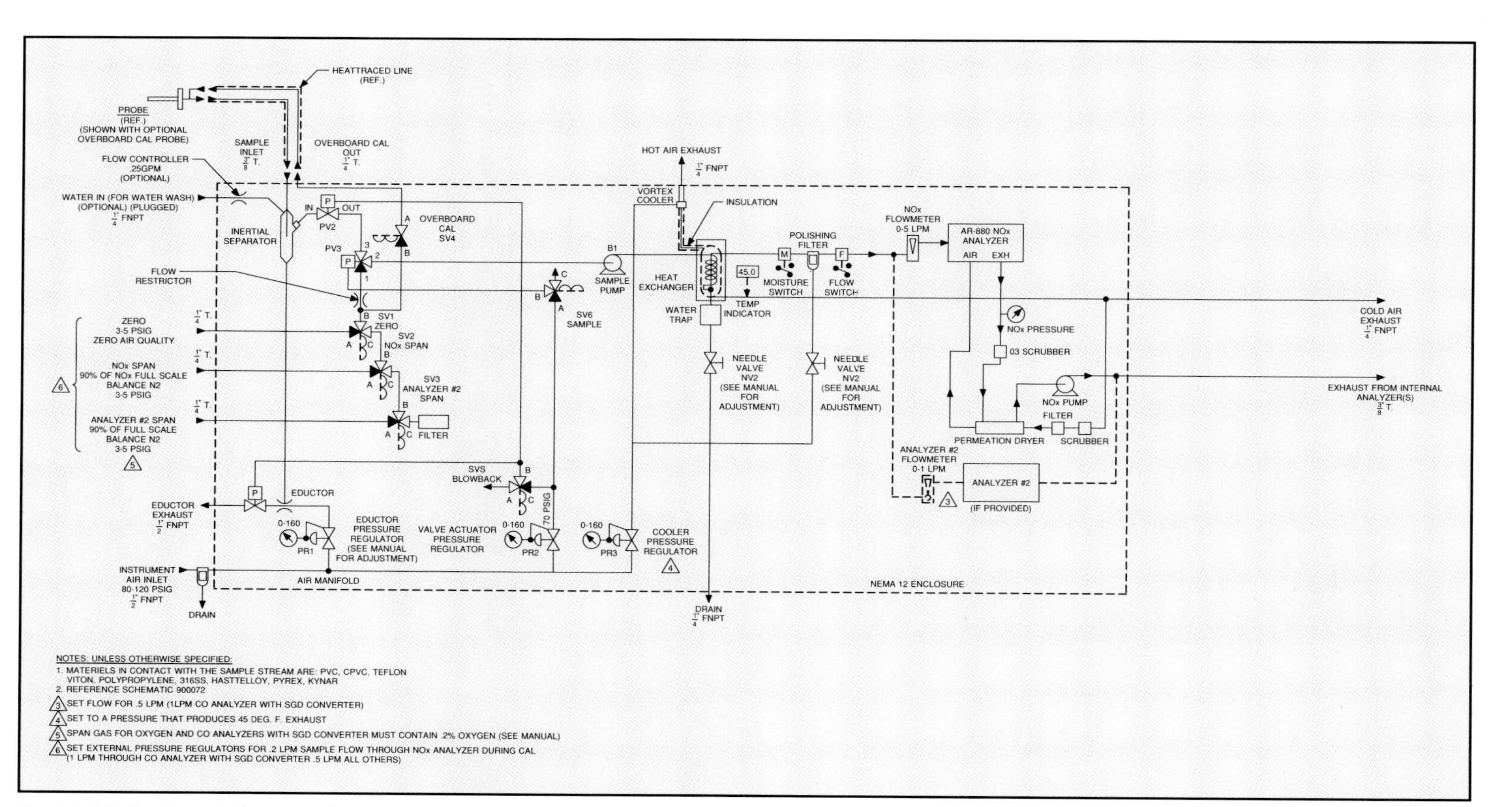

Fig. 2-18(a). Gas Sample Extraction System Diagram (Courtesy of Anarad, Inc.)

Fig. 2-18(b). Multiple Gas Sampling and Analysis System (Courtesy of Anarad, Inc.)

sample is drawn from the gas stream, washed, and cooled to a temperature below its dewpoint. The resulting condensate is removed, and the remaining ''dry'' gas is passed through a remote analyzer at a constant volumetric flow rate.

In the ''wet'' process, a sample is drawn from the gas stream, passed through a heated gas detection cell housing, and returned to the gas duct. Since the gas temperature never drops below its dewpoint, the gas sample retains any moisture that was present before sampling.

The extraction method offers the following advantages and disadvantages:

Advantages

A. Analyzer can be located in an easily accessible and environmentally safe location.

B. Generally easier to service.

Disadvantages

A. Greater potential risk of air infiltration into sample stream.

B. Potential risk of condensation in the sample line.

C. Analysis typically does not occur under the same conditions as those at the sampling point in gas stream (temperature, pressure, etc.).

Infrared Absorption

A relatively new technology for CEM is a sophisticated measurement technique known as infrared absorption (IA). This technique originated from, and is still used extensively for, laboratory analysis of sample gas constituents.

In the IA process, a beam of infrared light is emitted from a source unit and passes through the exhaust gas stream. It is then reflected back to a detector unit by a mirror assembly installed on the opposite side of the duct or stack. The infrared light source is chopped at a high frequency and synchronized with the detector in order to filter out stray light frequencies and noises that might be picked up by the detector (see Fig. 2-19).

An output signal from the detector assembly is transmitted to a remote analyzer, which then measures the amount of light absorbed by the gas stream at wavelengths corresponding to various gas concentrations (see Fig. 2-20).

Analytical wavelengths for several commonly measured gases are as shown in Table 2-4.

Compound	Wavelength (in micrometers)
Octane	3.4
Carbon monoxide	4.7
Nitric oxide	5.3
Sulfur dioxide	8.8
Methylene chloride	13.5

Table 2-4. Analytical Wavelengths of Common Gases

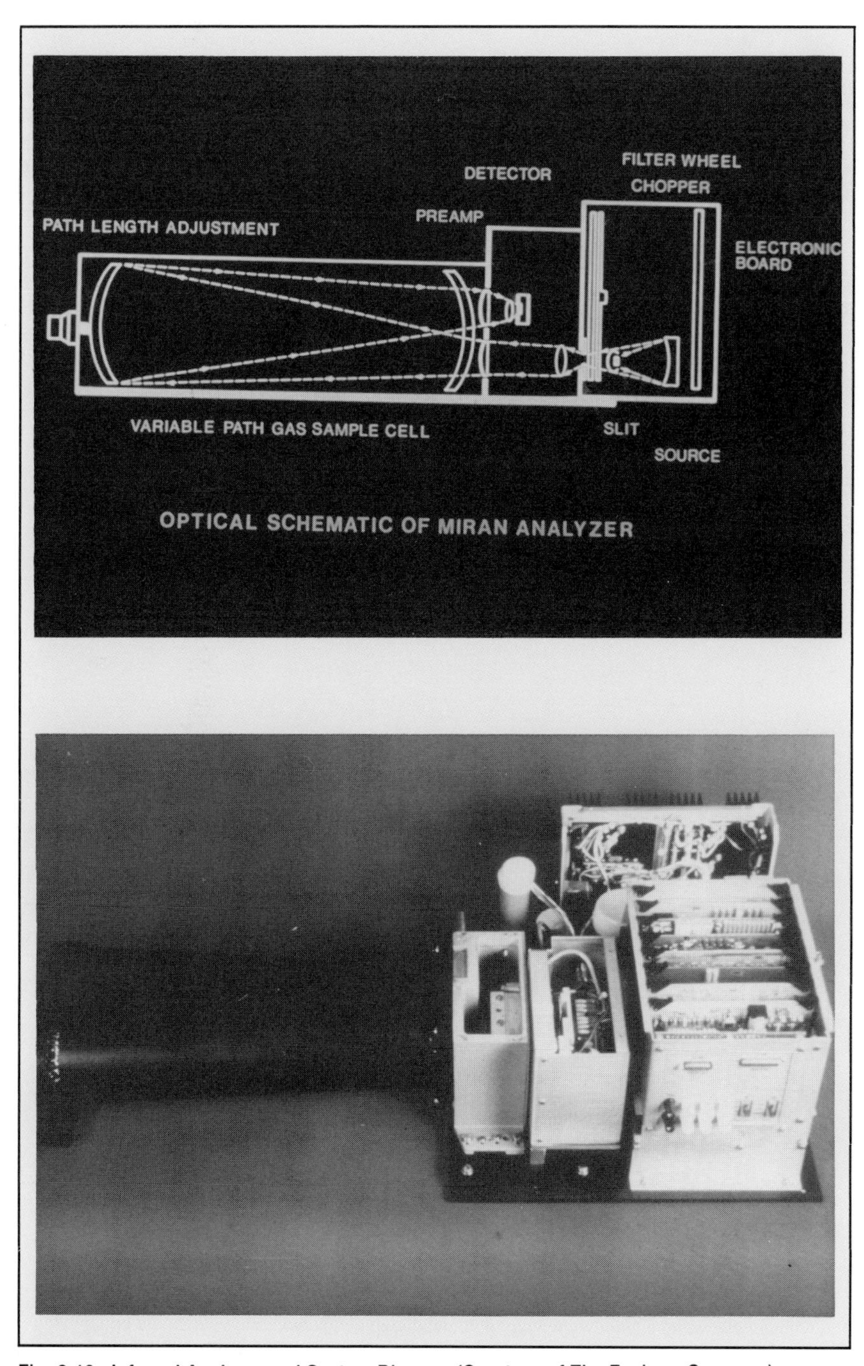

Fig. 2-19. Infrared Analyzer and System Diagram (Courtesy of The Foxboro Company)

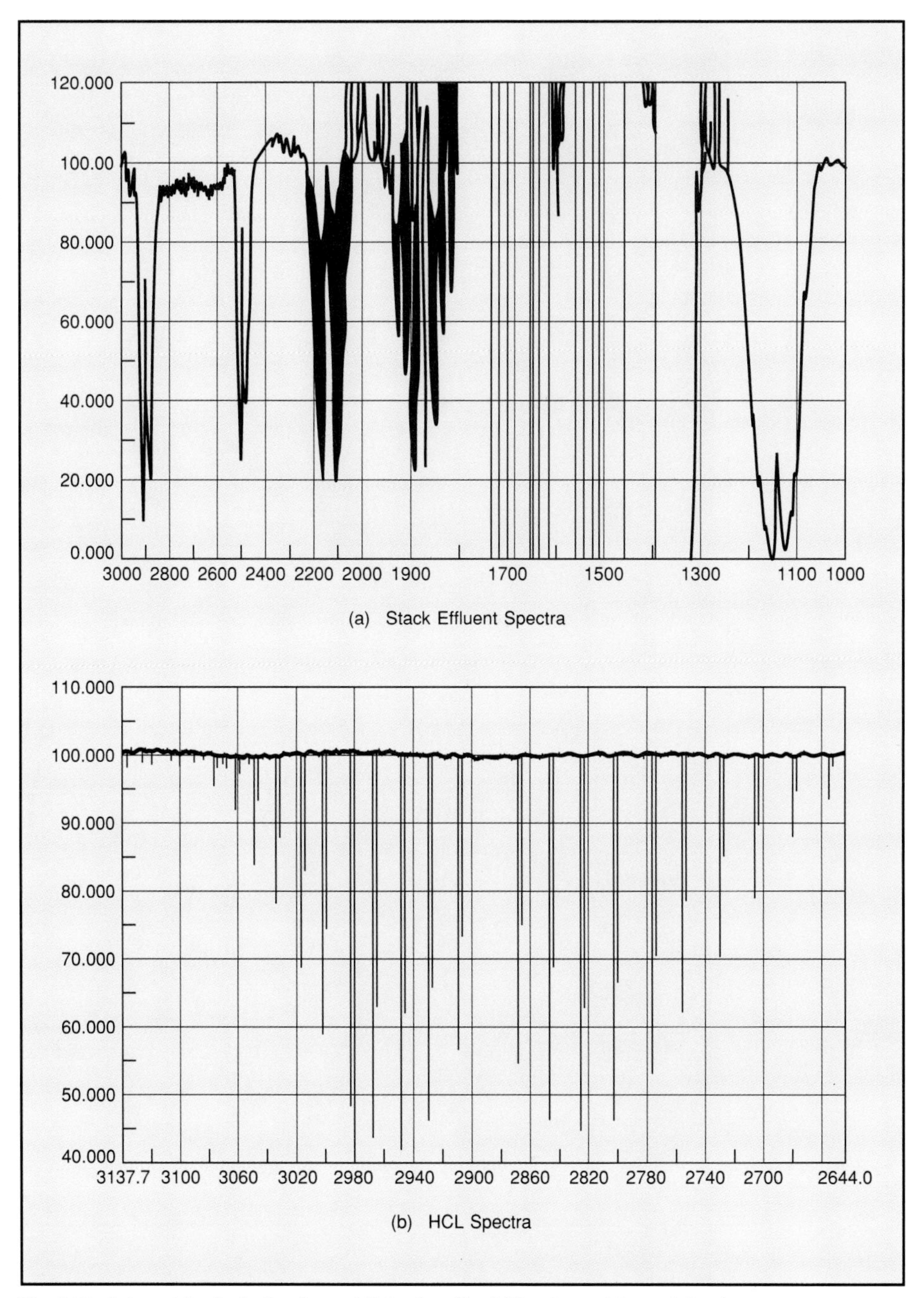

Fig. 2-20. Infrared Analysis Graphs and Selection Chart (Courtesy of Anarad, Inc.)

Accuracy of IA units is typically on the order of ±2% of full scale with a repeatability of ±1% of full scale.

Some analyzers of this type are capable of monitoring up to six different gases per sensor, including carbon monoxide, carbon dioxide, hydrocarbons, sulfur dioxide, hydrochloric acid,

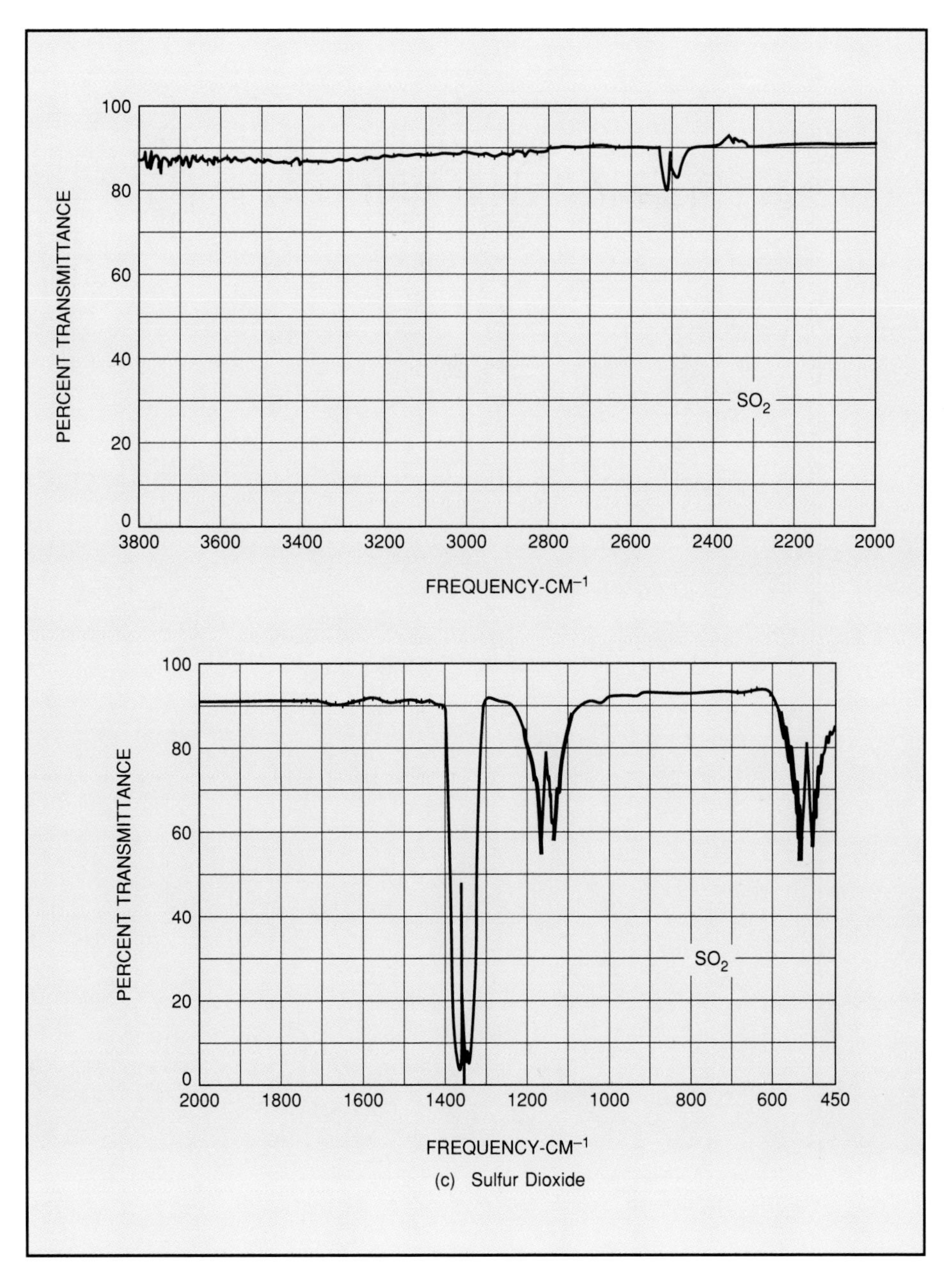

Fig. 2-20. (Continued)

hydrogen fluoride, nitrous oxide, ammonia, water, and opacity and particulates.

2-5. Dispersion Analysis

In order to establish the impact of stack emissions on the surrounding area (resulting ground level concentrations), it is necessary to perform a dispersion analysis.

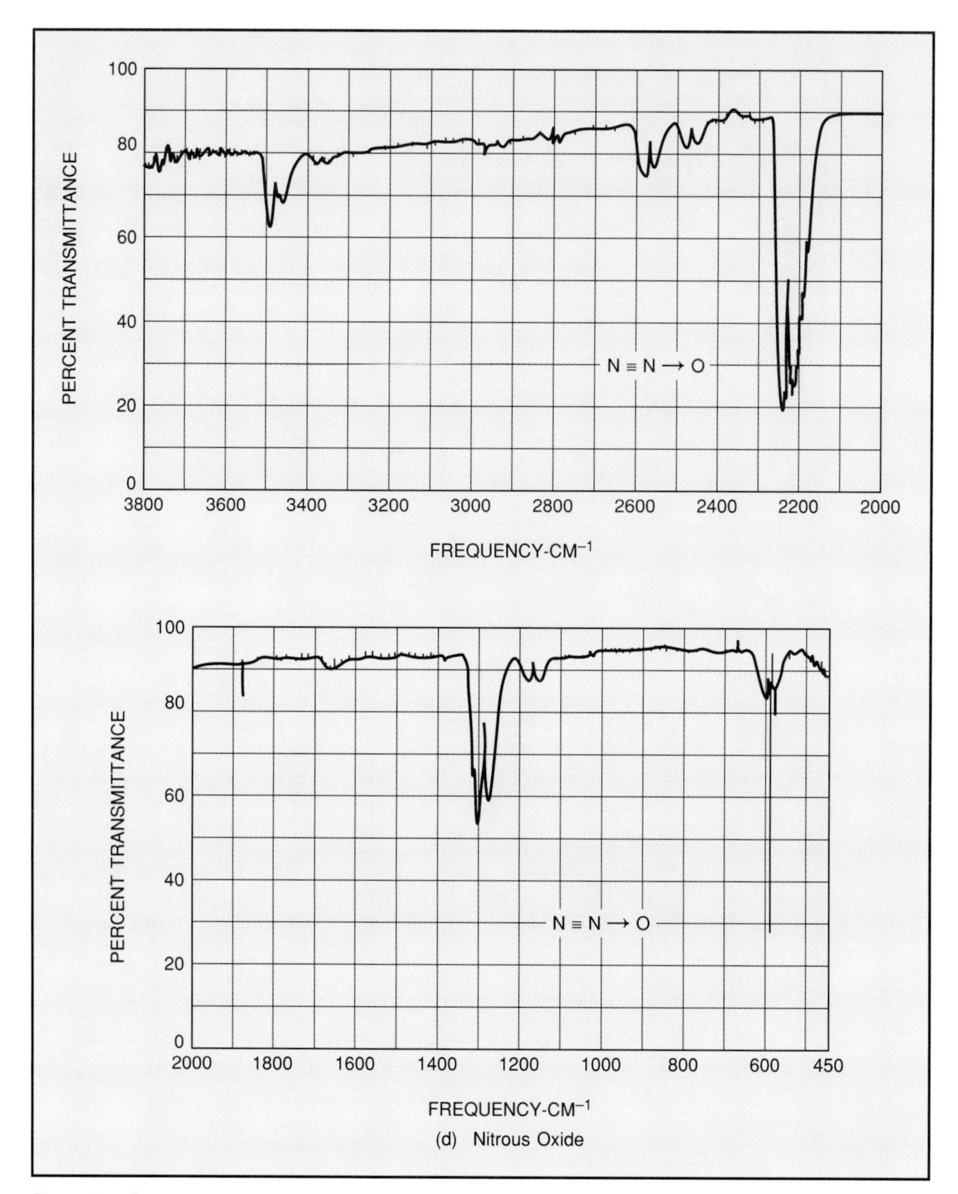

Fig. 2-20. (Continued)

Exhausted air and/or gas exits a stack or chimney at a velocity dependent upon the design of the fan system (brake horsepower, fan design, rpm, duct diameter, etc.), or the stack height, diameter, and gas temperature of a "natural draft" system.

The exit velocity of the exhausted air/gas mixture, in combination with stack height with relation to surrounding structures, determines the altitude it reaches before it begins to

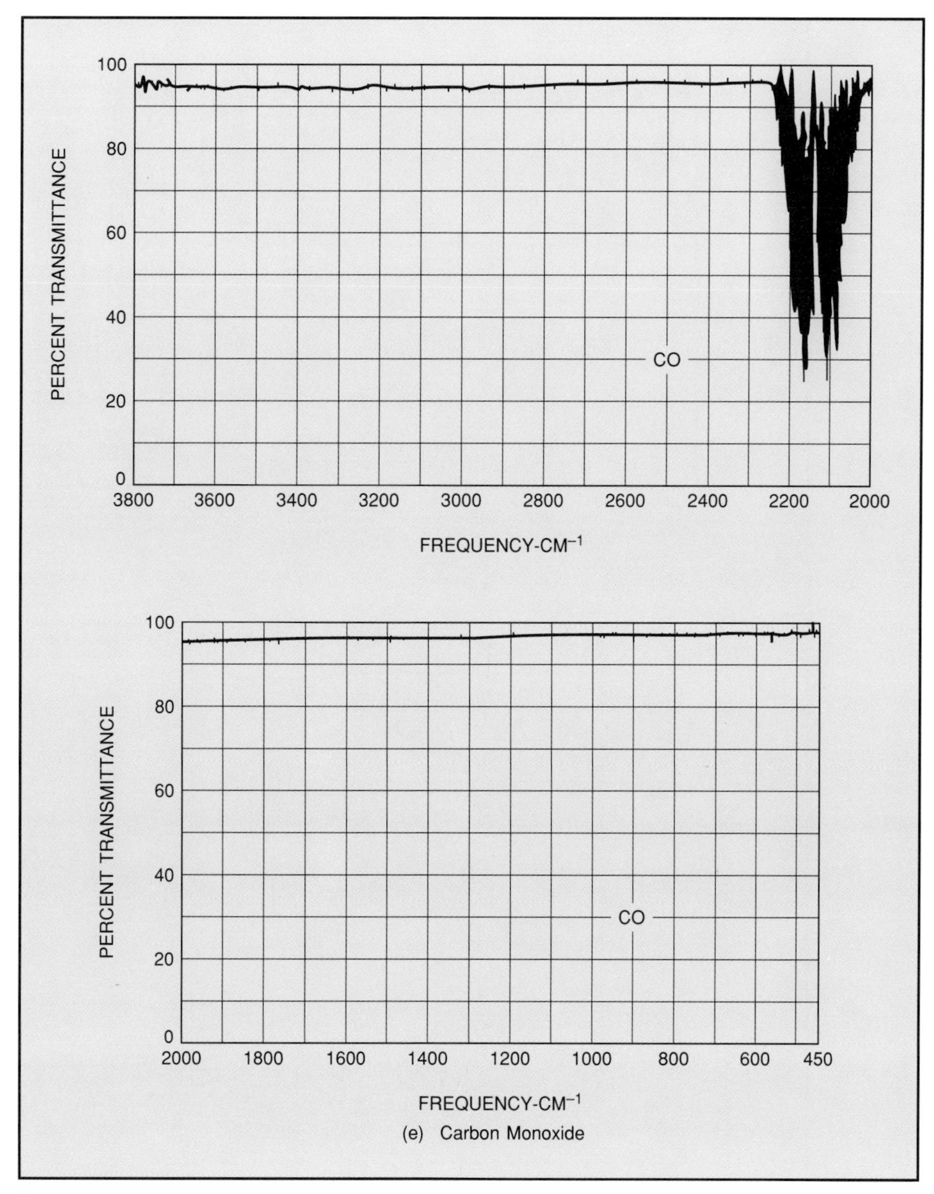

(e) Carbon Monoxide

Fig. 2-20. (Continued)

disperse into the atmosphere. The greater the exit velocity and/ or higher the stack height, the lower the measurable levels of pollutants at ground level (gases are dispersed into the atmosphere before they reach the ground).

Meteorological data needed to evaluate air quality impacts from a facility include hourly monitoring of wind direction and speed at the top of the stack, as well as air temperature and stability.

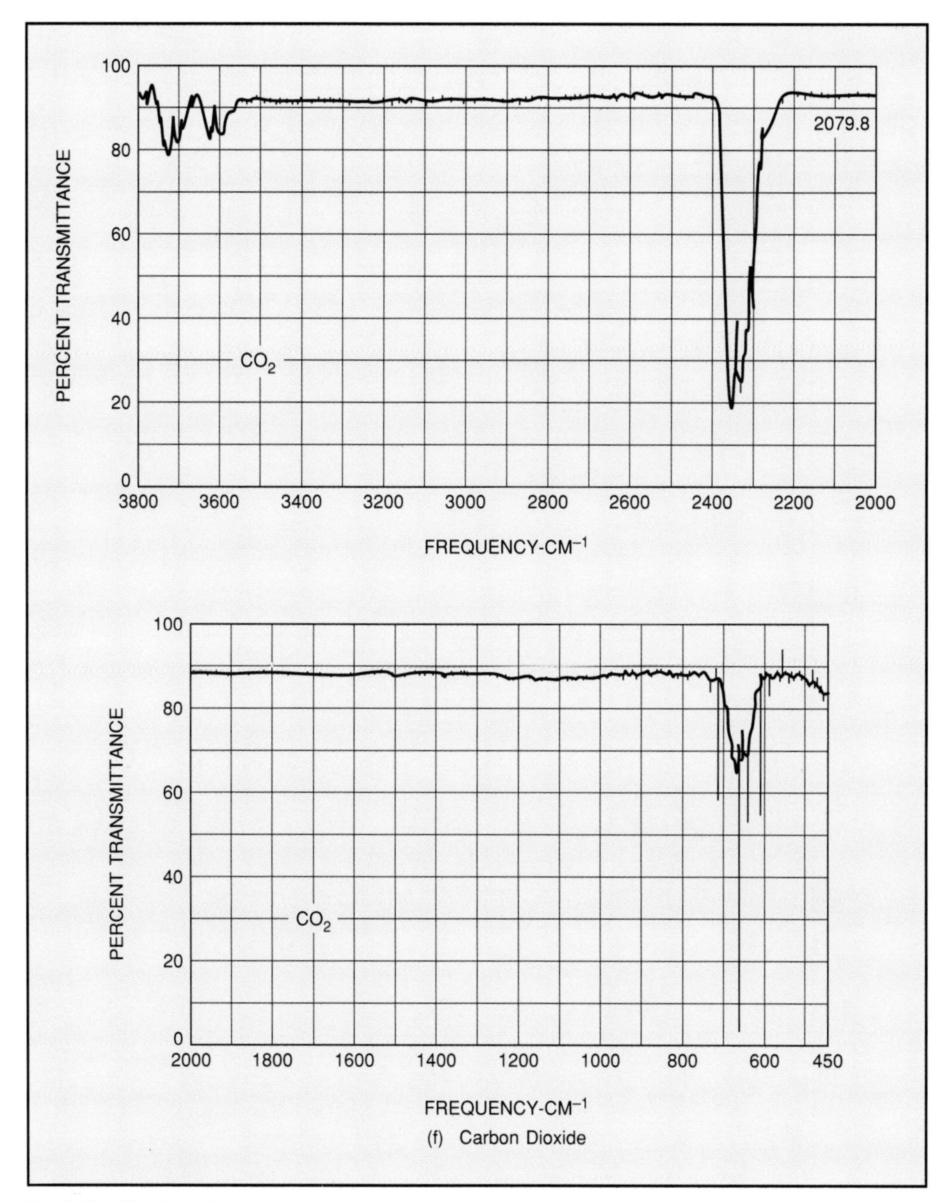

(f) Carbon Dioxide

Fig. 2-20. (Continued)

Point Source Screening Model

Typically, a version of EPA model PTMTP is used to estimate 1-hour average concentrations that occur during typical weather conditions.

A well-known Gaussian equation is used in estimating pollutant concentrations for a right-hand, three-dimensional

GAS MEASUREMENT	NONDISPERSIVE INFRARED	CHEMI-LUMINESCENCE	FLAME IONIZATION	THERMAL CONDUCTIVITY
CARBON MONOXIDE	YES			YES
CARBON DIOXIDE	YES			YES
NITRIC OXIDE	YES	YES		
NITROUS OXIDE	YES			
SULFUR DIOXIDE	YES			YES
TOTAL HYDROCARBONS	YES			
METHANE	YES		YES	
ETHYLENE	YES			
PROPANE	YES			
HEXANE	YES		YES	
FREON	YES		YES	
AMMONIA	YES			YES

(g) Gas Analyzer Selection Chart

Fig. 2-20. (Continued)

coordinate system with a single point source located at coordinate (0,0,0):

$$X_{(x,y,0:h)} = Q/(\pi\sigma_{y,z}U) * \exp(-1/2(y/\sigma_y)^2) * \exp(-1/2(h/\sigma_z)^2)$$

where:

$X_{(x,y,0:h)}$ = ground level pollutant concentration at point x,y for a stack height h in grams/cubic meter (g/m^3)
Q = pollutant emission rate (g/m^3)
σ_y, σ_z = standard deviation of the plume concentration distribution in the cross wind (y) and vertical (z) directions.
U = wind speed at stack height (m/sec)
y, z = standard deviation of the plume

Variables contained in this equation are either model inputs or are calculated based on data input to the model.

Estimating Effective Stack Height

Effective stack height is a function of the physical height of the stack, plume rise, and "downwash" effects.

Plume rise is calculated by using an equation recommended by Briggs (Ref. 5) for stable and unstable conditions:

$$h = 1.6F^{1/3}U^{-1}x^{2/3}, \quad x < 3.5x^*$$

where:

h = plume rise due to buoyancy (m)
U = wind speed at stack height (m/sec)
x = downwind distance to receptor (m)
F = buoyancy flux (m^4/sec^3)
$= gV_sR_s^2((T_s - T_a)/T_s)$
$x^* = 34F^{2/5}$ for $F > 55$
V_s = average stack velocity (m/sec)
R_s = inside radius of stack (m)
T_s = temperature of gas at stack (kelvins)
T_a = atmospheric temperature (kelvins)
g = acceleration due to gravity (m/sec^2)

Wind velocity at the stack is extrapolated from the value at the measurement location (typically a 10-meter tower) by assuming an exponential vertical profile in accordance with the USEPA (1977).

Downwash of the stack plume is caused by the aerodynamic effect of the stack according to the suggested Briggs (Ref. 5) equation:

Reduction in effective stack height due to downwash (m)
$= 4(1.5 - V_s/U)R_s$

where:

U = wind speed at stack height (m/sec)
V_s = average stack velocity (m/sec)
R_s = inside radius of stack (m)

For a more detailed explanation of dispersion coefficients and modeling data, see References 4, 7, and 8.

Computer simulation programs that model the dispersion of exhaust gases are available. These programs are capable of modeling the dispersion and projected ground level concentrations of multiple emission points within a defined geographical area.

Modeling Program Input Data

Data required as input for modeling programs includes:

A. effective stack height,

B. outside stack diameter,

C. inside stack diameter,

D. exiting gas temperature,

E. flow rate,

F. exit velocity, and

G. emission rate of the gas constituents at 100% (worst case) load condition.

Also included is meteorological data for the geographical location, such as:

A. wind direction,

B. wind speed,

C. stability,

D. atmospheric temperature,

E. atmospheric (barometric) pressure,

F. mixing height, and

G. receptor locations (coordinates and elevation).

The modeling program calculates the ground level pollutant concentrations for emission point load conditions of 100%,

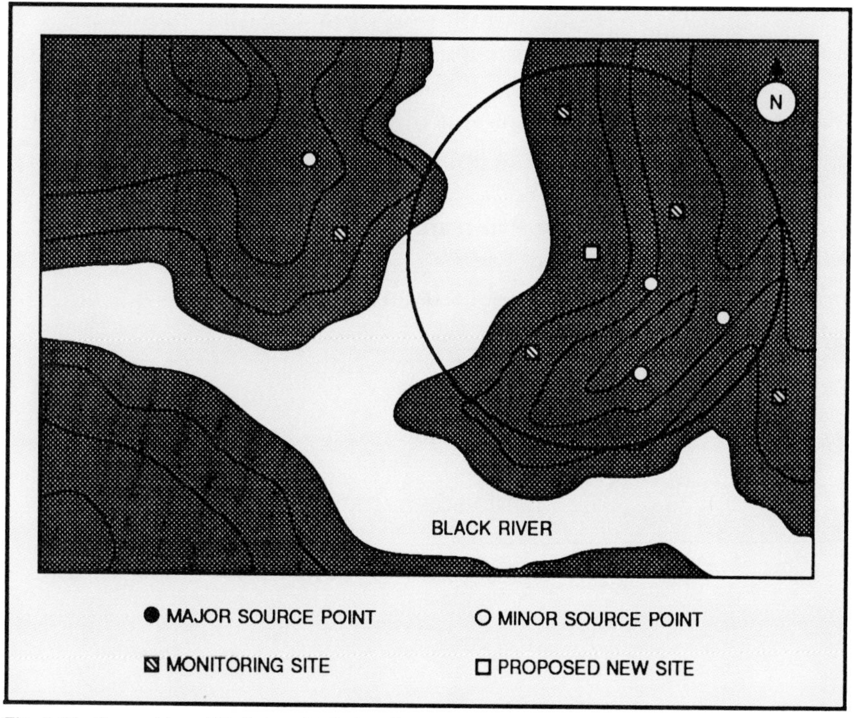

Fig. 2-21. Ground Level Pollutant Emission Control Monitoring

75%, 50%, and 25%. These concentrations are then extrapolated to give relative average concentrations for 8-hour, 12-hour, and annual periods.

The projected ground level concentrations are compared to permissible regulatory limits. If they are within an acceptable range, no further modification of the existing or proposed emission point is required. If ground level concentrations will exceed permissible limits, pollutant emission controls or system modifications that will bring those levels into an acceptable range must be provided (see Fig. 2-21).

2-6. Emission Control Systems

Precombustion Control of Sulfur Dioxide

An effective precombustion method to limit sulfur oxide emissions involves removal of sulfur from the fuel prior to its combustion. Sulfur content in coal is generally greater than in fuel oils (see Table 2-5).

Fuel Component	Bituminous Coal % by weight	No. 6 Fuel Oil % by weight	Natural Gas % by volume	Propane % by volume
Carbon	73.6	86.0	—	—
Hydrogen	5.3	11.0	—	—
Carbon monoxide	—	—	—	—
Methane	—	—	82.9	—
Ethane	—	—	14.9	2.2
Propane	—	—	—	97.3
Pentane	—	—	—	0.5
Illuminants	—	—	—	—
Oxygen	10.0	1.0	—	—
Nitrogen	1.7	0.2	2.2	—
Carbon dioxide	—			
Sulfur	0.8	0.8	—	—
Moisture	0.6	1.0	—	—
Ash	8.0			

Table 2-5. Fuel Analysis Chart

Many plants, particularly in the Northeast, have converted coal-fired units to burn oil or natural gas in order to reduce harmful emissions and to avoid the capital costs associated with coal burning and storage. These include leachate from coal piles exposed to rainfall and fugitive dust emissions created during the pulverization and conveying of coal from storage to the stokers. Coal-fired plants designed to overcome these obstacles generally burn a low-sulfur coal in order to help minimize SOx emissions.

Postcombustion Control of Sulfur Dioxide

Postcombustion removal of sulfur dioxides from stack gases is commonly referred to as flue gas desulfurization (FGD). FGD is accomplished by a variety of control methods, including the following.

Limestone Injection

An additive (calcium carbonate) is injected directly into the furnace of the boiler. Hot combustion gases cause the calcium carbonate to chemically react with sulfur dioxide to form calcium sulfate ($CaSO_3$) particles. The calcium sulfate is then removed by a particulate removal system. Sulfur dioxide removal efficiency using this method is typically between 20 and 40% (see Fig. 2-22).

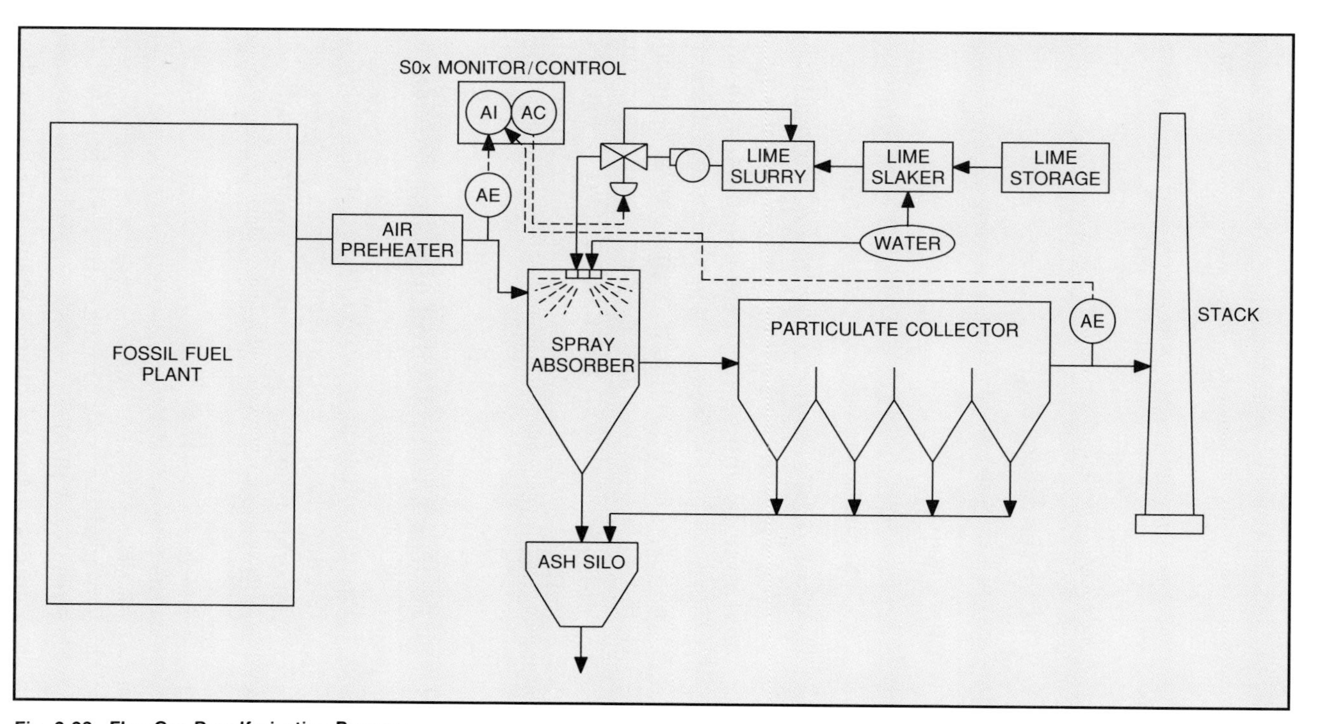

Fig. 2-22. Flue Gas Desulfurization Process

Wet Scrubbers

Calcium carbonate, calcium hydroxide, or magnesium oxide, in the form of a slurry (about 20% solids), is added to the scrubbing liquid (spray) to react with gaseous sulfur dioxide. Absorber systems such as this typically remove up to 85% of the sulfur dioxide (even higher removal efficiencies have been reported), which is then disposed of in the resulting wastewater. This system is also quite effective in removing particulates from the gas stream. Absorber vessels are sized to provide sufficient contact and reaction time of the sorbent spray with the sulfur dioxide, as well as time required for the spray droplets to evaporate.

While capital costs to install this type of system are relatively low, operating costs can potentially be high. Most of the operating expense can be attributed to the cost of sorbent chemical additives and transport and disposal of the large volume of wastewater (see Fig. 2-23).

Dry Sorbent Systems

Combustion gases pass through a fixed sorbent bed that contains materials that readily absorb sulfur dioxide, such as activated carbon. The sorbent material is then regenerated and used again. Sulfuric acid is removed as a by-product of this process (see Fig. 2-24).

Construction and Operating Costs

Capital and annualized costs must be considered when selecting the optimal emissions control system for a plant. Such costs, associated with maintaining the process, include maintenance, sorbent material supply, water, electrical power, disposal of waste materials, and spare parts.

In addition, the pollution control system will create additional pressure drop in the gas flow stream, which must be overcome by natural draft or induced draft fans to provide adequate escape velocity (i.e., plume height) from the stack.

Required plume height will be mandated by the existing background concentrations, average weather conditions, topography, applicable regulations, and dispersion analysis.

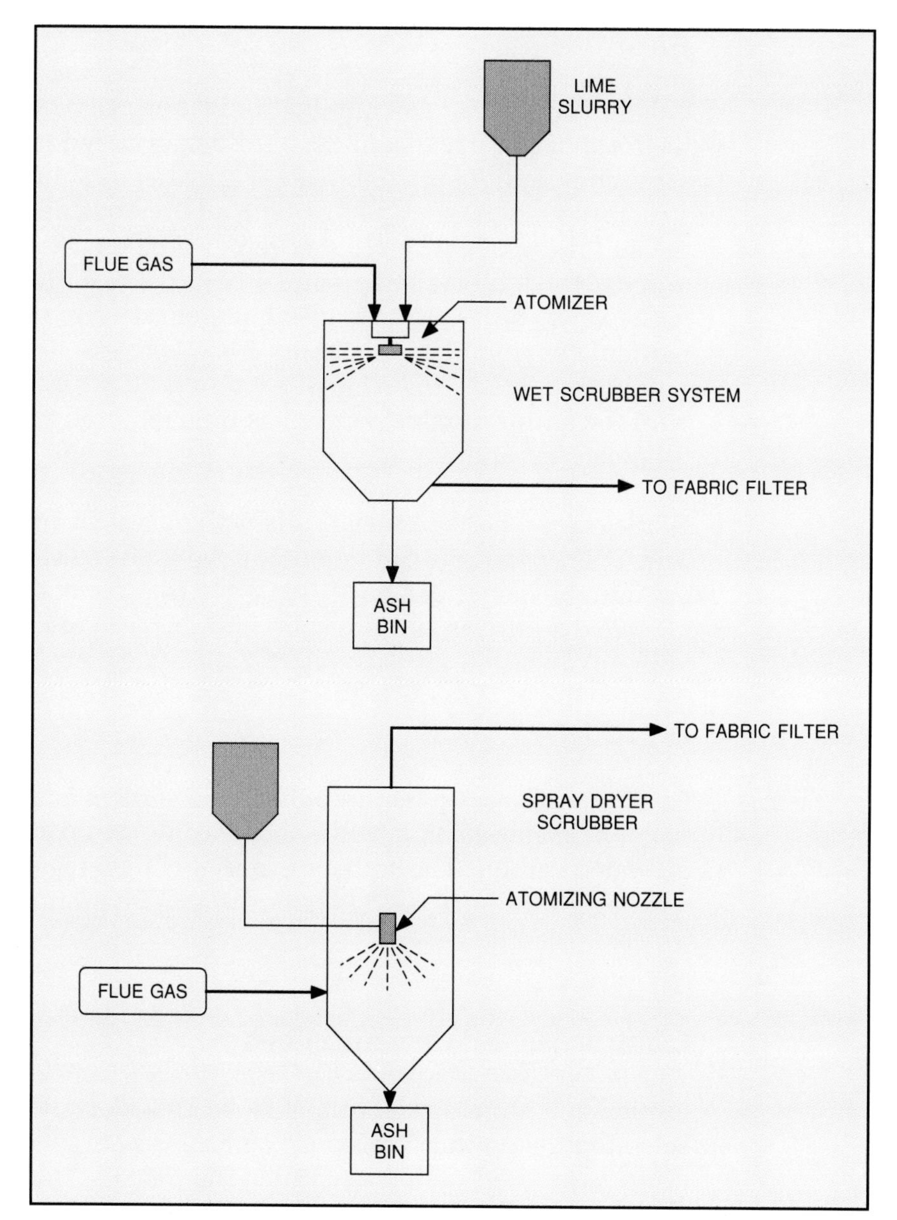

Fig. 2-23. Wet and Dry Scrubber Systems

This may result in a need for increased fan horsepower or chimney height.

Precombustion Control of Nitric Oxide

The two most popular forms of NOx control are modified combustion control and selective catalytic reduction (SCR).

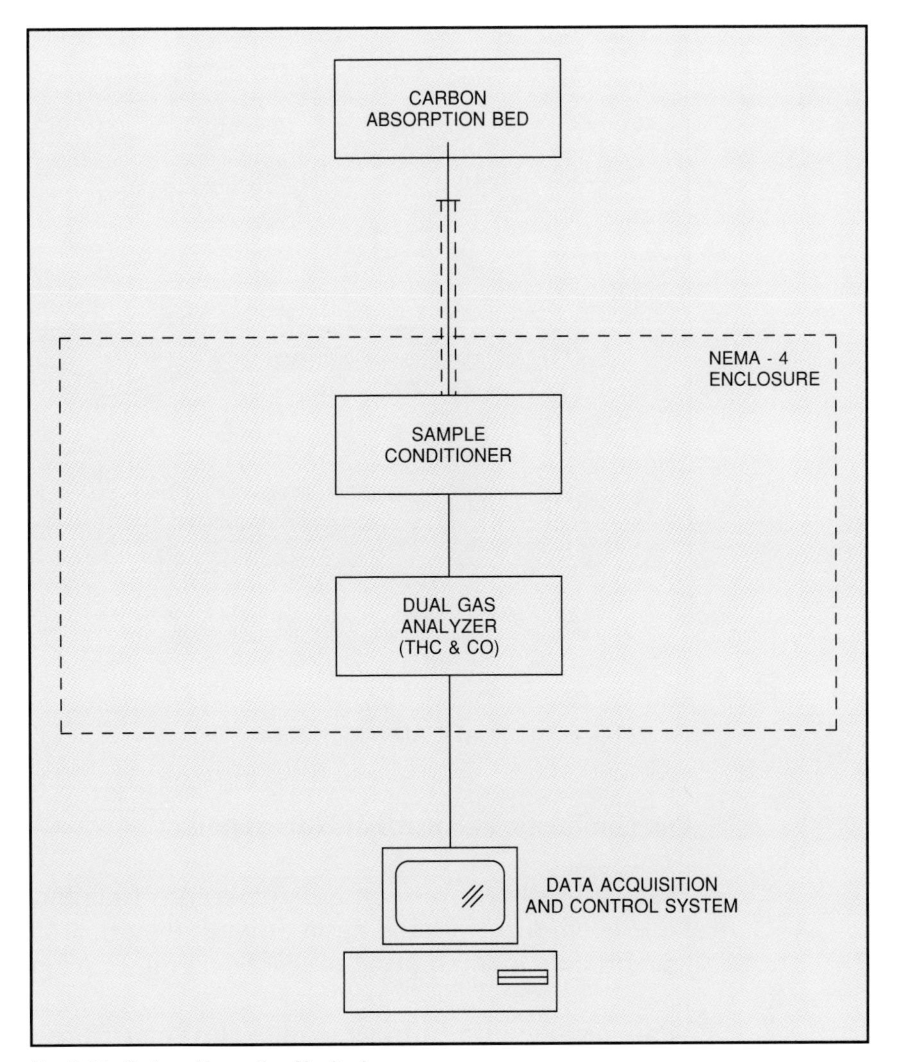

Fig. 2-24. Carbon Absorption Monitoring

By far the more common practice of NOx control is modified combustion to limit the maximum flame temperature. As stated earlier in this section, NOx production is significantly lower at reduced combustion temperatures.

Among the combustion control methods that reduce NOx production are modified fuel injection, water-emulsified fuel oil, combustion air temperature control, and precombustion (see Fig. 2-25).

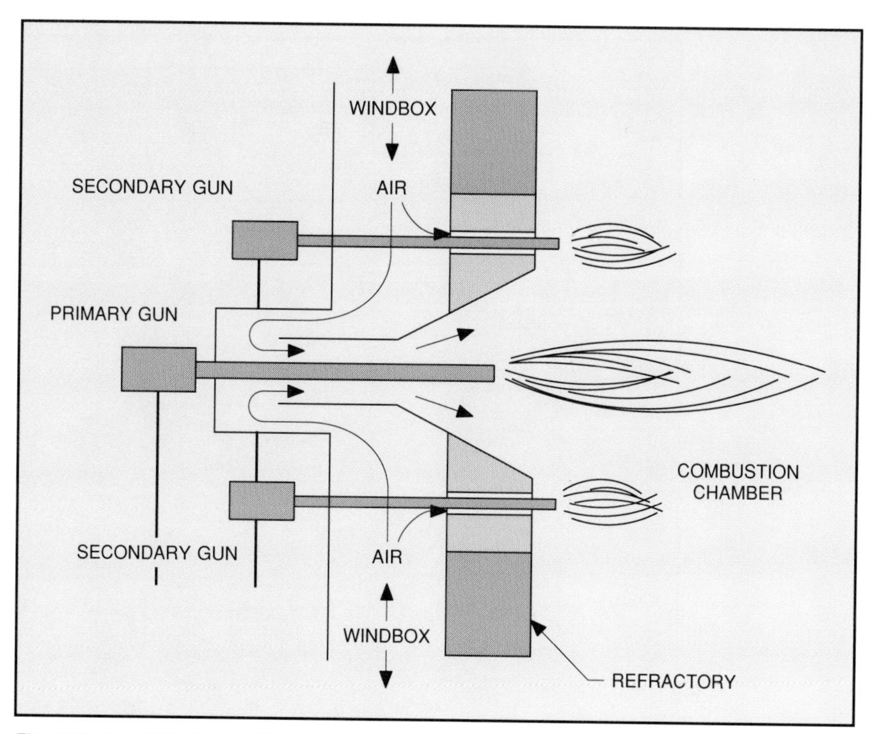

Fig. 2-25. Low NOx Burner (Fuel Staging Method)

Postcombustion Control of Nitric Oxide

Postcombustion control of NOx emissions from burning fossil fuels is often considered more difficult than that of other gas constituents in attempting to meet existing EPA limits.

Methods to control nitrous oxide (NOx) emissions include selective catalytic reduction (SCR). In this process, ammonia (NH_3) is injected into the high temperature flue gas stream where it reacts with the NOx in the presence of a metal catalyst (platinum) to form nitrogen (N_2) and water vapor, two chemicals harmless to our environment (see Fig. 2-26).

This method is very efficient in removing oxides of nitrogen from the gas stream (see Table 2-6). However, ammonia has been identified as a toxic substance under SARA (Superfund Amendments and Reauthorization Acts) Title 3. Therefore, it must be handled very carefully. In addition, a potential exists for unreacted ammonia to be emitted into the atmosphere (referred to in the industry as ''slip''). Therefore, precise control of ammonia injection is required.

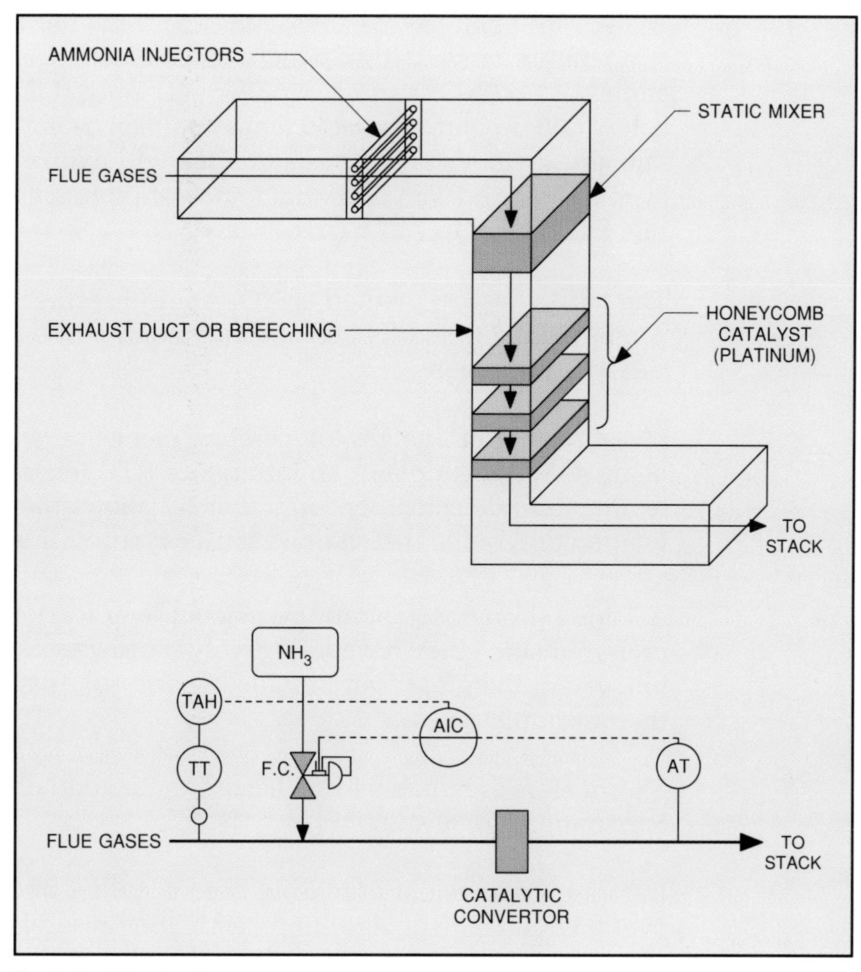

Fig. 2-26. Selective Catalytic Reduction with Ammonia Injection

Fossil Fuel Plants	Unit No.	Boiler Capacity, megawatts	Catalyst Design	Inlet NOx, ppm	Outlet NOx, ppm	NOx Removal, % efficiency	Unreacted Ammonia, ppm	Inlet SOx, ppm	Flyash, lb/ft^3
Takehara	1	250	Plate	350	67	81%	4	—	—
"	2	250	Tube	350	67	81%	4	—	—
"	3	700	Plate	250	48	81%	2	700	—
Shimonoseki	1	175	Honeycomb	420	180	57%	1	1000	0.001
Shin-Ube	1	156	Honeycomb	400	140	65%	2	—	—
Mizushima	1	156	Plate	350	120	65%	2	—	—
Saijo	1	156	Honeycomb	380	130	65%	1	—	—
"	2	250	Honeycomb	330	90	70%	1	—	—

Table 2-6. SCR Unit Performance

Another difficulty associated with SCR is the narrow gas temperature range within which the chemical reaction between ammonia and NOx can occur. Efficient removal does not occur below 600°F, and unreacted ammonia may be emitted in the flue gases. Above 800°F the ammonia will oxidize, creating even more NOx! The optimum temperature range required for safe SCR is 600 to 750 degrees.

The most effective method of NOx control seems to be a combination of modified combustion control *and* selective catalytic reduction.

Measurement of flue gas temperature can be accomplished using a thermocouple or an infrared or RTD (resistance temperature detector) sensor. Table 2-7 shows the effective temperature range that various sensor types will work within.

A Type J (iron-constantan) thermocouple or RTD is generally preferred since they have a relatively narrow measurement range and, therefore, can provide better resolution than other thermocouple types.

Thermistors are unable to withstand the potentially high gas temperatures (800–1200°F).

Infrared is capable of providing good accuracy and is effective over a wide range of conditions with good resolution. However,

Temperature Measurement Sensors

Sensor Range (°F)	Thermistor	RTD	Thermocouple	Infrared
−100 to 0	Yes	Yes	Yes	Yes
0 to 500	Yes	Yes	Yes	Yes
+500 to 1800	No	Yes	Yes	Yes
1800 to 4200	No	No	Yes	Yes
4200 to 6000	No	No	No	Yes

Thermocouple Ranges

Type & Material	Temperature Range (°F)
Type J (iron-constantan)	0 to 1700
Type K (chromel-alumel)	-50 to 2400
Type T (copper-constantan)	-300 to 700
Type E (chromel-constantan)	-100 to 1800
Type R (platinum-platinum)	0 to 2800

Table 2-7. Temperature Sensor Ranges

infrared detection tends to be much more expensive than RTDs or thermocouples.

Control of Particulates

Methods to control particulate levels in stack gases include the following.

Electrostatic Precipitators

Flue gases pass through an electrostatic field created by a series of charged electrodes (see Fig. 2-27). As particulates in the flue gas pass through the charged field, they become charged with the same polarity. Continuing on, the particles then become attracted to metallic plates with the opposite electrical charge. As a result, particulates cling to the collector plates until power to the system is interrupted and ''rappers'' or sonic horns are activated to shake the particles loose. The particles then drop and are collected in hoppers at the base of the unit for later disposal.

Since there is little constriction of gas flow through this type of filter system, the pressure drop across the unit due to gas flow restriction is relatively low. A large pressure drop could require increased induced draft fan horsepower, or greater stack height on a natural draft system, in order to attain the desired gas flow rate.

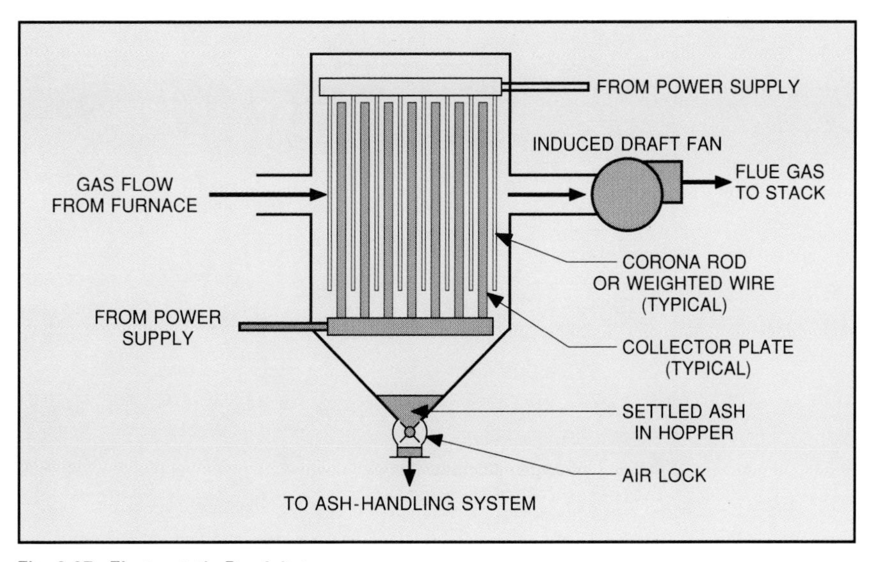

Fig. 2-27. Electrostatic Precipitator

Fabric Filters

Flue gases pass through a series of bag-shaped temperature and chemical-resistant woven fabric filters (see Fig. 2-28). The filter bags have a low air-to-cloth ratio (typically between 1.5 and 2.0) and are constructed of woven fiberglass or a combination of fiberglass and Teflon™ weave. As larger particles become trapped between the weave of the fabric, the efficiency, or capture rate, of the smaller diameter particles increases. High-pressure air or sonic horns are used periodically to shake the particles loose. The particulate, which has become ''caked'' onto the outside of the bags, drops into collection hoppers where it is collected for later disposal.

Scrubber Systems

Flue gases pass through a high-pressure liquid spray system, absorbent packing material, or a combination of the two. Particles are knocked out of the flue gas stream as they are struck by water molecules in the spray. They then drop to the bottom of a hopper where they are collected for later disposal.

There are literally hundreds of scrubber configurations. Most are custom designed based on gas composition, volumetric flow rates of the gas, spacial limitations, and operating environment.

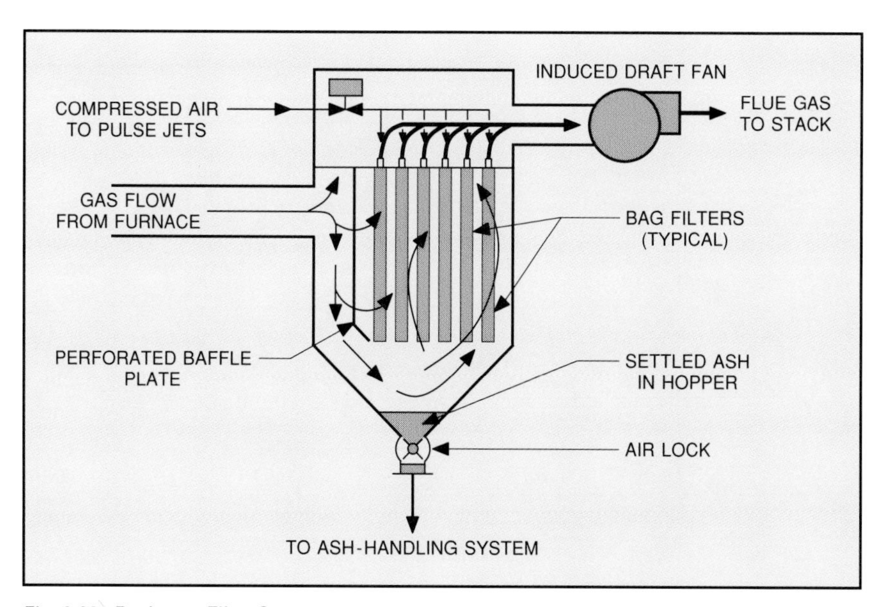

Fig. 2-28. Baghouse Filter System

Cyclones

The flue gas stream enters a chamber at a high velocity where it is forced into a downward spiraling cyclone pattern and then returns back up through the center of the chamber and out to the stack. Centrifugal force causes airborne particulates in the flue gas to strike the outer wall of the chamber during its spiralling motion and fall to the bottom where the residue is collected and disposed of (see Fig. 2-29).

Control of Incinerator Emissions

Growing concern over contamination of groundwater has rapidly forced the closing of many landfills throughout the United States. This move has created a resurgence in the process of incineration of waste materials to produce useable energy. The volume of residual waste to be disposed of at landfills is greatly reduced, while at the same time the heat produced by incineration is used to generate steam. The steam is used for heating purposes or to create electrical power by using the steam to turn a turbine-driven generator. Occasionally the steam is used for both purposes. In such a case, the process is referred to as "cogeneration."

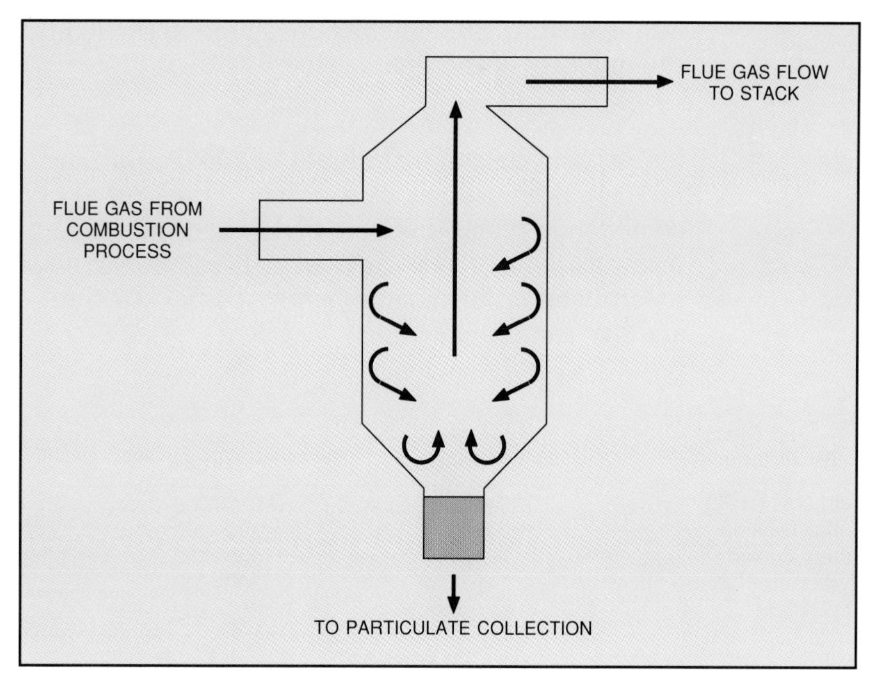

Fig. 2-29. Cyclone Diagram

Challenging problems develop when one is trying to maintain control of toxic stack emissions that result from burning waste materials. Products of incomplete combustion include heavy metals (lead, mercury, cadmium) and trace organic compounds (dioxins and furans); and acid-gas by-products such as sulfur dioxide, nitric oxide, hydrogen chloride, hydrogen fluoride, and nitric oxide, along with particulates, are produced during burning and are found to some degree in the flue gases (see Table 2-8).

Heavy metals such as lead and mercury vaporize at high flame temperatures. As the metal vapor begins to cool, it condenses and attaches itself to airborne particulates, a large percentage of which are collected by particulate control systems such as baghouses or ESPs. It then becomes part of the ash material to be disposed of.

Since heavy metals and other toxic substances are often found in the remaining bottom ash and fly ash after incineration, safe disposal of the ash also becomes a concern. This waste material must be transported to designated ash landfills equipped with containment liners that prevent eventual leaching of the toxic materials into surrounding groundwater. Monitoring wells are established in strategic locations around the landfill to monitor the integrity of the liner system.

Many basic control techniques employed for the control of incinerator emissions are similiar to those used to control combustion emissions from utility and industrial boilers. These methods include such devices as fabric filters, electrostatic precipitators (ESPs), scrubbers, cyclones, catalytic converters, and dry-sorbent injection.

Flue Gas Air Pollutant	Typical Range, ppm	Gas Temperature, °F	Gas Flow Rate, acfm
Acid Gas (HCL)	300–1200	300–500	13,000–250,000
Sulfur Dioxide	100–500	300–500	13,000–250,000
Sulfur Trioxide	2–15	300–500	13,000–250,000
Hydrogen Fluoride	5–25	300–500	13,000–250,000
Particulate	0.04–10 (gr/scf)	300–500	13,000–250,000

Table 2-8. Typical Waste-Burning Plant Flue Gas Emissions

Combustion Control

Proper control of combustion is very important to the control of pollutant emissions (see Fig. 2-30). By optimizing combustion of raw materials, the products of incomplete combustion are kept at a minimum level. Additional incentive is created by efficient energy production (Btu's of heat energy per unit volume of fuel burned) that is experienced during total combustion.

Optimization of combustion to minimize the level of PIC emissions involves a certain amount of compromise. Generally, combustion at the higher flame temperatures needed to guarantee destruction of PICs results in a greater level of NOx emissions. Therefore, a temperature of combustion that will effectively produce the desired low levels of both is desirable.

It has been determined that combustion conditions that produce low levels of carbon monoxide (CO) also result in more thorough destruction of organic compounds. Therefore, CO

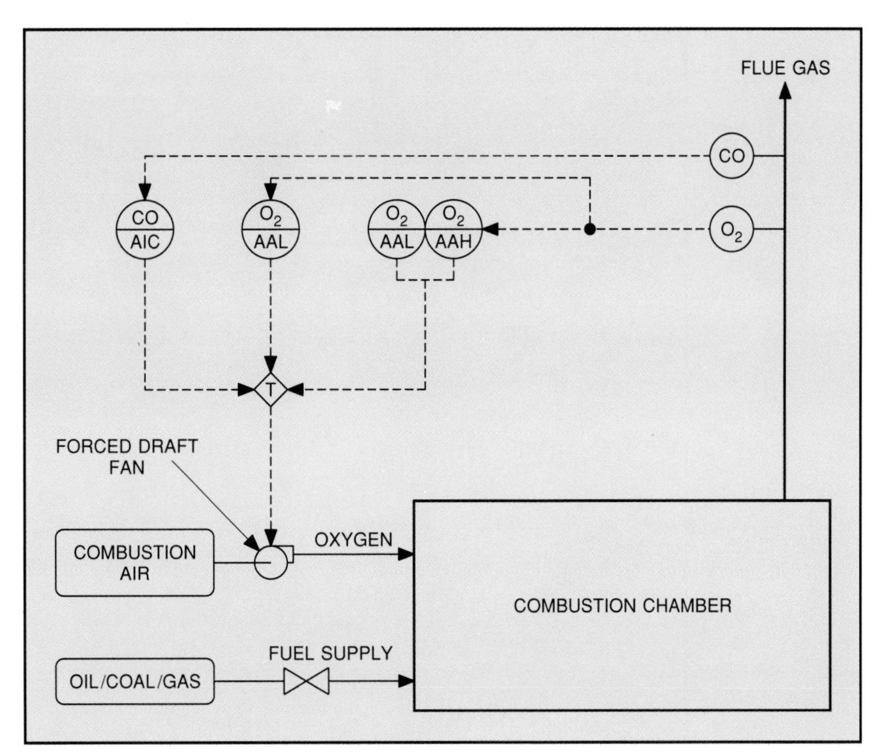

Fig. 2-30. Carbon Monoxide Closed-Loop Control

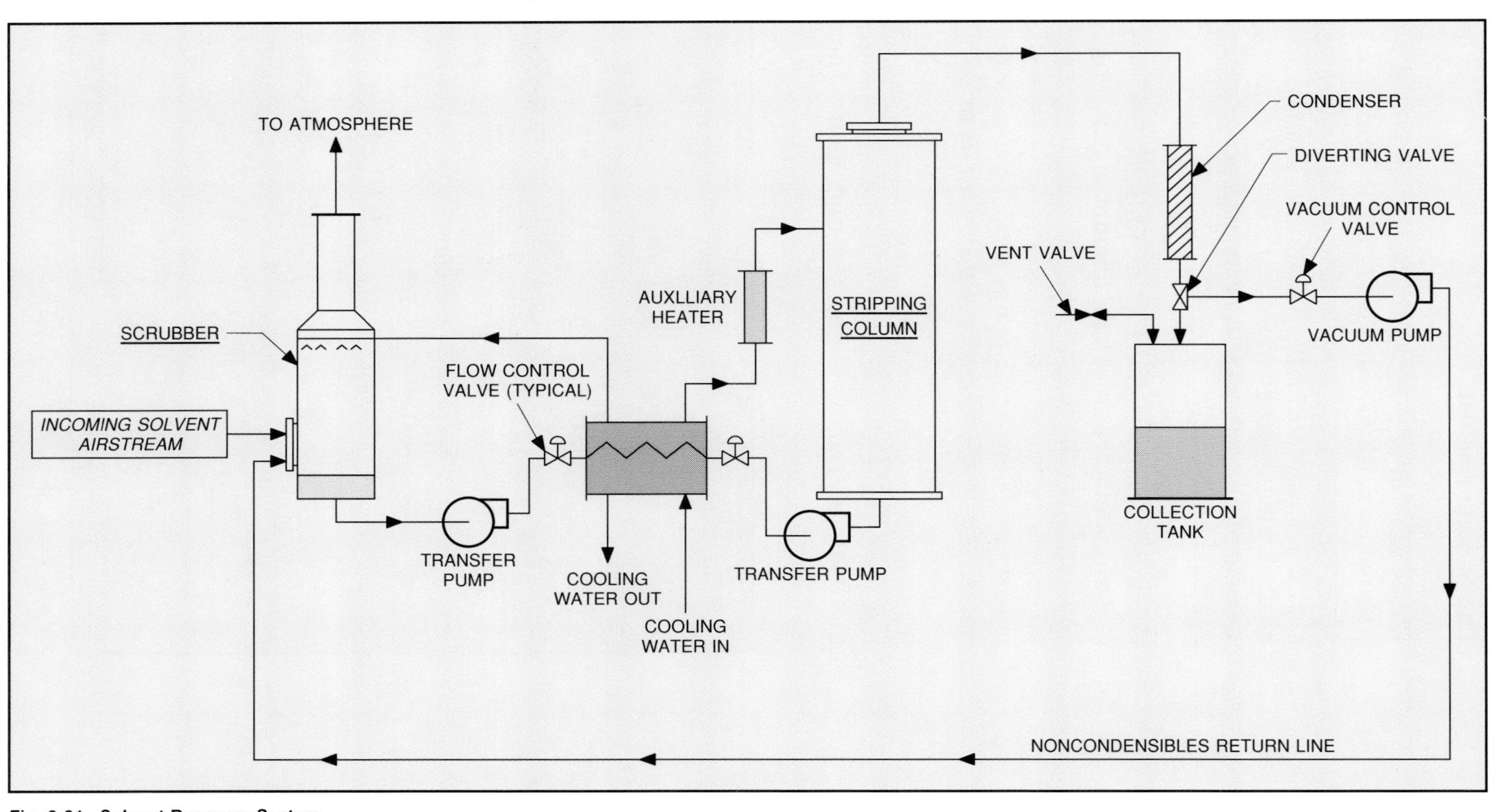

Fig. 2-31. Solvent Recovery System

monitoring becomes a useful tool in controlling the emission of organic compounds. It should be noted, however, that in applications requiring high levels of excess air, the use of CO monitoring to trim combustion control can be difficult.

Another effective approach in the reduction of waste-burner emissions is to use source separation of the fuel prior to combustion. Lead, cadmium, NOx, and CO emissions can be reduced by separating source materials out of the waste fuel supply, which is known to contain these substances. While source separation can reduce levels of these toxic substances, it is technically very difficult to ''clean'' raw waste materials to such a level that scrubbers or other such emission control devices can be eliminated from the pollution control process. Automating the separation process presents a somewhat complex and costly operation.

2-7. Chemical Recovery Systems

An ideal solution to controlling emission levels of air contaminants would be to recover the emitted chemicals or useful compounds so they could be reused or sold. Revenue from the recovered materials would help defray the cost of operating the pollution control system.

Successfully recovering the chemicals first requires collection or capture of the controlled substance from the exhaust gas stream. This is typically accomplished by some type of filtering method or chemical reaction that separates the material from other components of the gas stream (see Fig. 2-31).

The captured substance must then be separated from any other materials used in the recovery process, such as filter media or chemical constituents. This is typically accomplished by condensing the material out using a heating and refrigeration system.

References

1. ''Experiences with CO Measurement in Industrial Boilers,'' *Instrumentation in the Power Industry*, Instrument Society of America (1987).
2. *Anarad Inc. Analytical Instrumentation Catalog and Applications Guide*, Anarad, Inc. (1989).
3. McIlvaine, Robert W.,''Control Technology for MSW Incinerator Applications,'' Paper presented at the 79th Annual Meeting of the Air Pollution Control Association (APCA) (1986).

4. *Valley User's Model Guide*, Publication No. EPA-450/2-77-018, Office of Air Quality Planning and Standards, NC: EPA (1977).
5. Briggs, Gary A., *Some Recent Analyses of Plume Rise Observations*, ATDL File Contribution, NOAA (1977).
6. "Plume Rise Predictions," *Lectures on Air Pollution and Environmental Impact Analyses*, Boston: American Meteorological Society (1977).
7. *Workbook on Atmospheric Dispersion Estimates*, Publication No. AP-26, Research Triangle Park, N.C.: Environmental Protection Agency (1970).
8. *User's Guide to the Interactive Versions of Three Point Source Dispersion Programs: PTMAX, PTDIS, and PTMTP*, Research Triangle Park, N.C.: Environmental Protection Agency (1973).

Exercises

2-1. *In what engineering units are air emissions typically measured or described in terms of EPA limits?*

2-2. *What is the EPA's definition of a "major source" of airborne pollutants?*

2-3. *Identify three methods of limiting the amount of SOx emitted from a boiler plant or waste-burning facility.*

2-4. *What can be done to limit fugitive emissions from process control systems in industrial facilities?*

2-5. *Sketch the basic components of low-leakage valve packings. Describe their purpose. Are they totally effective in eliminating fugitive emissions?*

2-6. *Identify three basic types of gasket designs that are effective in reducing or eliminating fugitive emissions.*

2-7. *Identify and describe the basic types of analyzers that might be used to monitor flue gases at a waste-burning facility.*

2-8. *Sketch both a simple in situ system and an extraction-type gas sampling system for flue gas analysis. List several advantages and disadvantages of each.*

2-9. *What is dispersion analysis and what is its purpose? What type of instrumentation might be used to collect data for this type of analysis?*

2-10. *Identify the types of pollution control systems used to limit SOx, NOx, and particulates. How effective are these systems in removing contaminants? How would you monitor their performance?*

2-11. *Identify and describe the basic components of an FGD system. Sketch the system, including process controls and instrumentation required in order to work properly.*

2-12. *Describe the purpose of ammonia injection for the purpose of SCR (selective catalytic reduction). How efficient is this type of system for removing NOx from a gas stream? Identify potential problems with trying to control this type of system.*

2-13. *Explain the difference between precombustion and postcombustion control and cite an example of each method.*

Unit 3:
Indoor Air Quality

UNIT 3

Indoor Air Quality

This unit addresses the indoor environment in which we live and work. Indoor air quality (IAQ) can have a more immediate impact on health and welfare than the outdoor environment. While nature offers some assistance in dampening the immediate impact of pollution, the effects of poor air quality control practice indoors are confined to the relatively small, enclosed areas we frequently occupy.

OSHA regulation 29 CFR 1910.94 requires employers to use appropriate instruments, procedures, or environmental controls to maintain the recommended quality of air in the workplace.

Learning Objectives — When you have completed this unit, you should:

A. Be aware of the typical sources of poor indoor air quality.

B. Know the types of instrumentation available for portable and continuous analysis of ambient air quality.

C. Be familiar with instrumentation and control systems used to maintain proper indoor air quality.

D. Understand relative humidity measurement and control and its importance in IAQ control.

E. Be familiar with building systems design practices that help prevent poor IAQ.

3-1. Government Regulations

The Occupational Health and Safety Administration (OSHA) has created a list of exposure levels for gas concentrations that have been determined to pose a potential health hazard to people or are suspected carcinogens. Refer to Appendix D.

While few of these substances are typically found in commercial facilities, many may be present in an industrial

Degreasing Areas	Fumigation Facilities
Methylene chloride Trichloroethylene Tetrachloroethane Perchloroethylene Naptha	Methyl bromide Ethylene oxide Acrylonitrile Ethylene dybromide Ethylene dichloride
Polymer Manufacturing	**Pesticide Manufacturing**
1,3-Butadiene Styrene Chlorotrifluoroethylene Ethylene Vinyl chloride Phosgene	Methyl isocyanate Triethylamine Methylene chloride
	Refineries
	Benzene Toluene Xylene Ethyl benzene

Table 3-1. Potentially Hazardous Gas or Vapor Compounds (Source: The Foxboro Company)

facility or a process plant. A potential exists for worker exposure to any type of material used in or stored at a process facility.

Table 3-1 lists examples of potentially hazardous gas or vapor compounds typically found and monitored for OSHA compliance. Methods to detect and measure these gases or vapors are identified later in this unit.

3-2. Causes of "Sick Building Syndrome"

Ventilation

Within the past two decades, a concern for energy conservation measures to limit rising utility costs has led to building designs that significantly reduce the amount of outdoor air that can infiltrate through cracks and crevices around windows, doors, and other exterior building components. Fresh air entering the building is limited to that required for minimum ventilation.

People who occupy space within a building breathe in available oxygen and exhale carbon dioxide. Oxygen within the space is replenished by introducing more fresh air. The American Society of Heating, Refrigeration, and Air-Conditioning Engineers (ASHRAE) recommends that the minimum volume of outdoor air entering a building for ventilation purposes be 15-20 cfm per occupant. This quantity is based, in part, on the fact that the average adult, during normal activity, respirates about one cubic meter of air per hour.

Note that the term ''outdoor air'' is used in lieu of ''fresh air.'' Depending upon the location of the facility, outdoor air may not be what is generally considered ''fresh'' (refer to Unit 2). The background levels of air pollutants determine the air's ''freshness.''

Minimum ventilation, combined with more efficient building insulations, has effectively reduced the heating or cooling energy needed to maintain a comfortable environment within most buildings. A ''comfort'' zone for human occupancy generally ranges between 68 and 78 degrees Fahrenheit. This range takes into account a wide variation in the physical activity that often takes place within a building.

Unfortunately, some materials commonly found in buildings exhibit ''outgassing,'' which is the emission of gaseous fumes or odors into the surrounding area. Sources of outgassing include vapors from volatile organic compounds (VOCs) such as some types of plastics, formaldehyde from wall and floor coverings, and ozone from photocopiers. Cleaning products can also be a source of outgassing. When present in high enough concentrations, these fumes result in eye and nasal irritation. It is often quite noticeable in newly constructed buildings once the interior finishes are applied and equipment is moved in. After the ventilation systems are activated for a few days, these concentrations diminish significantly.

The level of irritation varies with the gas concentration, the sensitivity of the people affected, and the level of exposure. When 20 percent or more of a building's occupants begin to suffer from symptoms of insufficient ventilation to limit the effects of outgassing and other sources of fumes, the facility is considered to be suffering from what is referred to as ''tight building syndrome.''

ASHRAE Standard 62-1989, ''Ventilation for Acceptable Indoor Air Quality,'' (Ref.2) identifies the minimum ventilation requirements for the health and comfort of human occupants in commercial facilities.

Also identified within ASHRAE Standard 62-1989 are guidelines for common indoor air pollutants within the industrial workplace (see Table 3-2).

Outdoor air requirements for adequate ventilating of industrial facilities are also based on building occupancy. Recommended industrial ventilation levels can be found by referring to procedures presented in 1986 Industrial Ventilation—A Manual of Recommended Practice, 1986 edition, American Conference of Governmental Industrial Hygienists (ACGIH).

While increased ventilation rates result in greater dilution of gas and vapor concentrations within a building, in most cases this air must be conditioned to meet temperature and humidity requirements before it enters the space. The more air to be conditioned, the greater the energy consumption and associated cost. Therefore, energy efficient alternatives should be investigated and perhaps a compromise considered. Any design compromises that can risk human health or welfare should not be considered.

It should also be mentioned that whenever outdoor air is brought into a building for ventilation purposes, the location of outdoor air intake louvers should be closely examined. More often than one would think, intakes are installed within several feet (sometimes inches) of exhaust duct outlets, resulting in toxic exhaust fumes being drawn back into occupied areas of the building. Occasionally, "fresh air" intakes have been installed through a wall bordering an alley or loading area that introduced carbon monoxide fumes into the building each time a truck delivered or idled near the intake.

Substance	Recommended Limits
Asbestos fiber	0.2–2.0 fibers/cc over 8-hr TWA
Carbon monoxide (CO)	50 ppm over 8-hr TWA
Formaldehyde	1 ppm over 8-hr TWA
Lead dust and fumes	0.15 mg/m^3 over 8-hr TWA
Nitrogen dioxide	3 ppm over 8-hr TWA
Ozone	0.2 ppm over 8-hr TWA
Sulfur dioxide	2 ppm over 8-hr TWA
[TWA = time weighted average]	

Table 3-2. ASHRAE Industrial Workplace Guidelines for Common Indoor Pollutants (Source: ASHRAE Standard 62-1989)

Humidity

Another potential source of poor indoor air quality is a condition of extreme humidity. Relative humidity greater than 70% within a building is conducive to the growth of bacterial organisms, viruses, fungi, and dust mites. Mildew is a good example of a bacterial organism that develops rapidly in a humid area. Growth of these organisms is further enhanced by the presence of materials within the building that contain high percentages of cellulose, a primary food source. Such materials include fiberboard, lint, dust, and dander.

Occupant illnesses related to airborne microorganisms fall under a perhaps broader classification of poor indoor air quality, commonly referred to as ''sick building syndrome.''

Microbial contamination within a building is occasionally traced to stagnant water within HVAC (heating, ventilation, and air conditioning) systems and cooling towers. Chemically treating or boiling standing or stagnant water can be used to kill, or prevent the growth of, unwanted microorganisms. One well-publicized effect from this type of contamination is known as ''Legionnaire's Disease.''

Potable (treated drinking) water should always be used to supply ''cold water'' space humidifiers. Untreated, vapor or mist created by this type of humidifier can potentially distribute airborne microbial bacteria present in the water supply.

Water used to create steam or water vapor for humidification purposes that directly contacts the air stream and has been chemically treated with toxic additives, such as amines for prevention of mineral scaling in boilers, can create a potential health hazard. Toxicity levels of any water treatment chemical additive must be examined. Alternative methods may avoid potentially serious health risks. In facilities where direct steam or water vapor humidification is inappropriate, electrical humidifiers that operate by heating the water in a pan, using submerged electrical heating elements to produce steam vapor, can provide needed humidity. A steam coil submerged in the water pan can also be used to indirectly produce steam vapor.

ASHRAE Standard 55-1981 (Ref.7) identifies an indoor air relative humidity of 20–60% as an acceptable range for human comfort and health. An "optimum" control range for health and comfort is generally considered to be 40–60% RH (see Fig. 3-1).

High humidity conditions within a building can often be attributed to high outdoor air humidity conditions and a lack of adequate humidity control within the building. Unconditioned outdoor air entering a building by mechanical ventilation or infiltration can introduce excessive moisture into the space.

Another source of high humidity is an inadequate vapor barrier in the exterior walls, foundation, and roof of the building. Standard concrete block, wood, and latex or enamel paints are porous materials that can allow moisture to eventually migrate through and equalize conditions on either side of the partition if allowed. A thin vapor barrier, often consisting of a plastic film, paint or chemical sealant, will prevent migration.

It is important to remember that moisture always tends to migrate from a moist area to a drier area (like a sponge or a wick). Humidity (grains of moisture contained within a volume of air) will always attempt to equalize throughout a defined area.

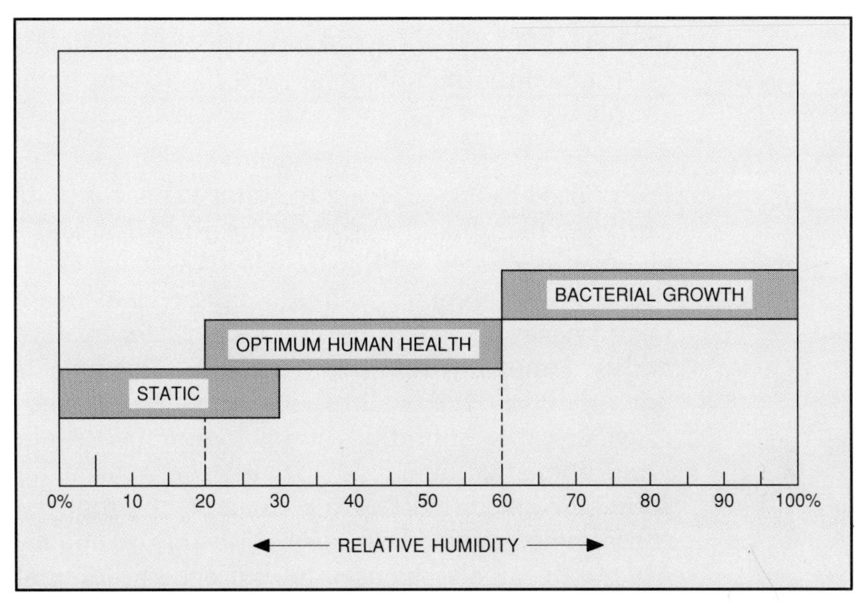

Fig. 3-1. Relative Humidity Range

Processes within a building may also produce water vapor as a by-product. Steam kettles, open vessels, liquid baths, and open sprays can also contribute significantly to high humidity conditions. Moist air exhaled by people as they breathe and perspiration contribute moisture to the environment.

Air Distribution Systems

Another potential contributor to the effects of tight building syndrome is an advanced method of mechanically distributing and controlling conditioned air called VAV (variable air volume). This method varies the volume of conditioned air distributed to occupied areas based upon their temperature requirements. By utilizing variable speed drives on the prime air movers (blowers), volumetric air flow is varied by modulating blower speed. Only the minimal energy (i.e., blower speed) required to maintain a comfortable temperature within the space is used (see Fig. 3-2).

If designed improperly, this energy efficient method of commercial air distribution can potentially result in a very low quantity of fresh air delivered to the space.

Duty cycling of HVAC equipment, an energy management technique whereby designated mechanical systems are periodically shut off for short durations to conserve electrical energy, has also contributed to poor air quality in some buildings (see Fig. 3-3).

Air becomes more stagnant the longer the ''off'' period, since no air flow or ventilation occurs during that time. Duty cycling also increases mechanical wear of HVAC and auxilliary equipment, resulting in shorter equipment life and increased maintenance. Therefore, duty cycling has been replaced in most applications by alternative methods of reducing energy consumption (variable volumetric flow, for example).

Conclusion

Poor indoor air quality can result in occupant discomfort, potential illnesses (such as sinusitis, bronchitis, and asthma), reduced worker productivity, and potential liability associated with related health effects.

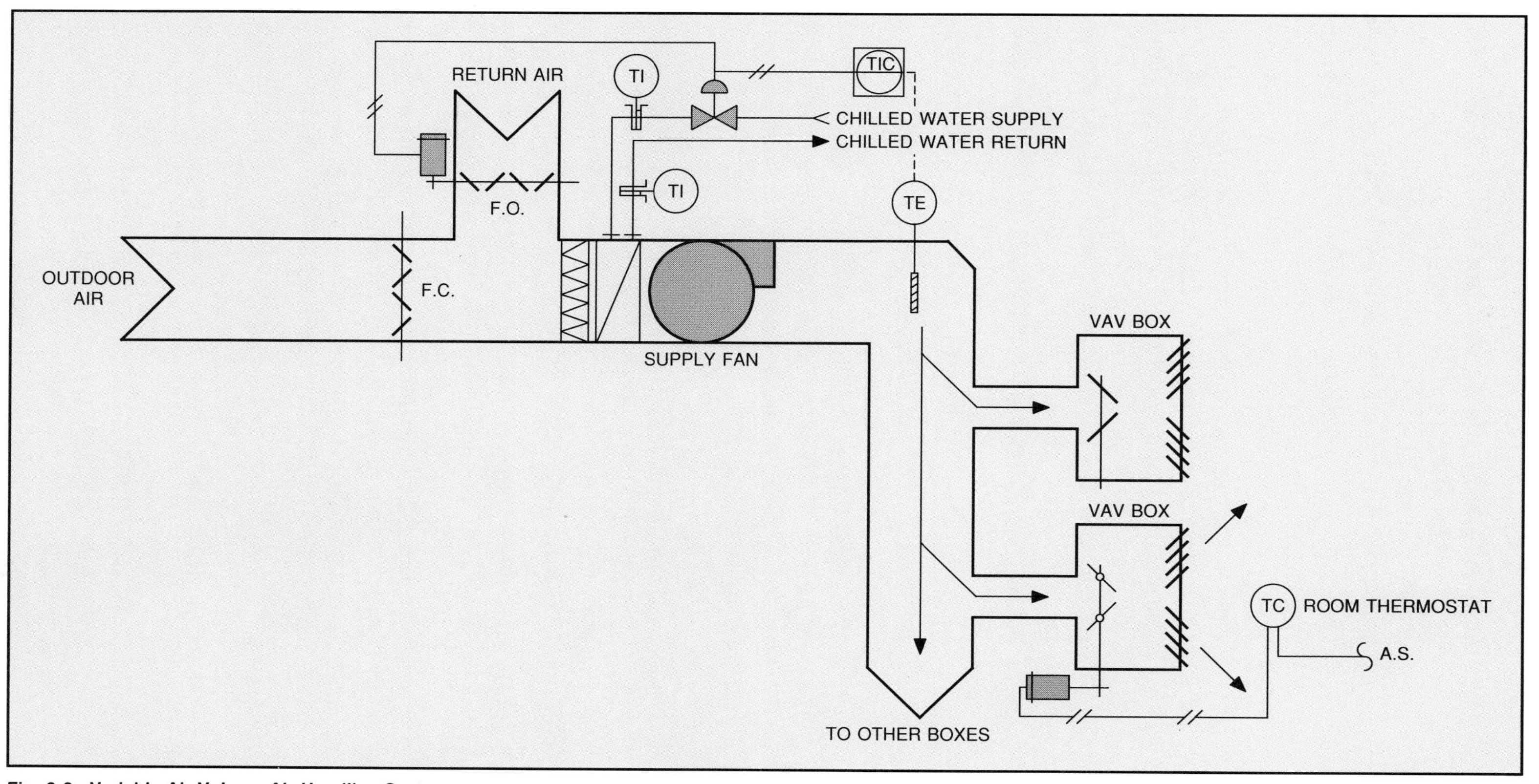

Fig. 3-2. Variable Air Volume Air-Handling System

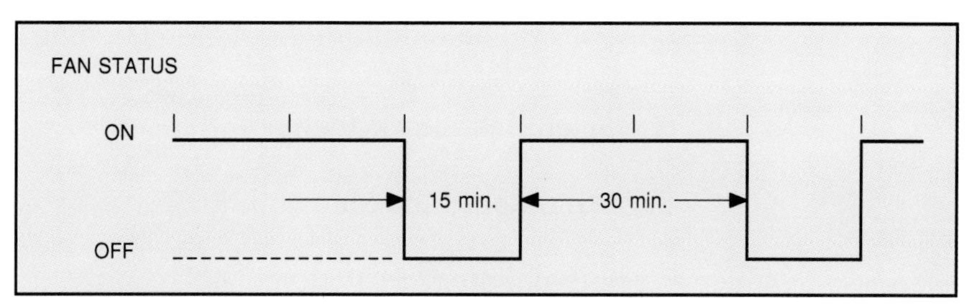

Fig. 3-3. Duty Cycling

3-3. Air Quality Sensors/Analyzers

Extensive numbers of analyzers are commercially available to measure the presence of air contaminants and assist in locating their sources. Many are designed to identify and measure a specific type of airborne pollutant. Others are capable of quantifying several different air pollutant types simultaneously. A number of these analyzers are designed for portable use and can also be used for continuous air monitoring. Various types of analyzers and their methods of operation are defined in this section.

Humidity Sensing

A variety of humidity sensing methods have been developed for applications ranging from laboratory analysis to industrial processes to commercial energy management systems (see Fig. 3-4).

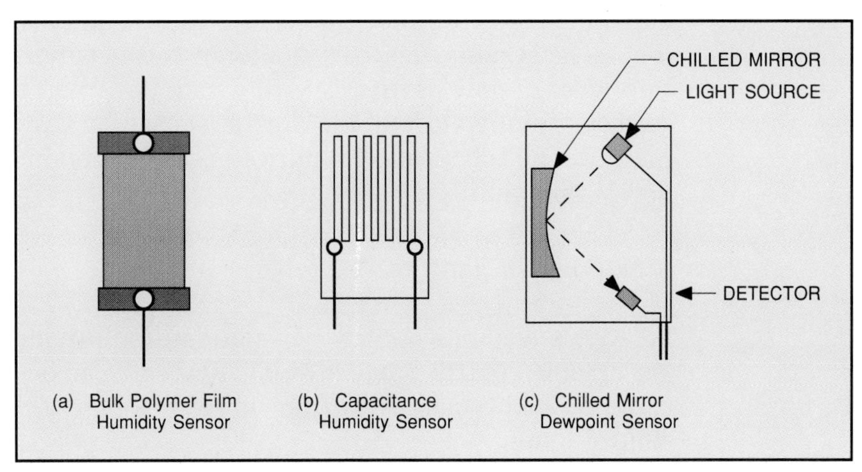

Fig. 3-4. Humidity/Dewpoint Sensor Designs

Examples of available humidity sensor technology include:

A. the chilled mirror (dewpoint),

B. lithium chloride (dewpoint),

C. bulk polymers (%RH), and

D. the psychrometer (%RH).

The accuracy of these devices range from a ±0.1°C dewpoint measurement using a chilled mirror, to ±3% RH with a bulk polymer. A typical sensible range is 15 to 99% relative humidity.

Sensing elements used in humidistats, which activate a snap-acting switch when a preset relative humidity limit has been reached, are constructed from materials such as bi-wood and human hair.

Chilled Mirror Sensors

Chilled mirror hygrometers provide a continuous dewpoint measurement by cooling a small mirror to dewpoint using a solid-state heat pump. An LED/phototransistor pair, bouncing an optical signal off the mirror, determines that moisture is forming on the mirror's reflective surface. Condensed moisture on the mirror's surface scatters light that would normally be reflected by a dry (room temperature) mirror, resulting in a lower signal at the phototransistor. The phototransistor's output is then amplified and serves as part of a rate feedback loop. The system stabilizes at a temperature condition where a thin layer of dew or frost is barely maintained on the mirror's surface. Temperature of the mirror (sensed by a precision platinum RTD beneath the mirror surface) is maintained at this dewpoint condition.

Bulk Polymer Sensors

Relatively accurate control of a standard building environment (acceptable control range 40–60% RH) can also be achieved using bulk polymer sensor/transmitters as part of the humidity control loop.

Bulk polymer sensors consist of a thin, moisture absorbent, plastic film impregnated with electrically conductive carbon. As humidity conditions change and the film absorbs more moisture from the air, physical and conductive properties of the film material change.

This change is amplified and characterized to provide an output signal (generally 4-20 mA DC or 1-10 V DC) proportional to the change in humidity. This type of sensor/transmitter arrangement is particularly effective in areas where humidity conditions change slowly, such as in a room. Unless the air changes in a room occur quite rapidly or serious control problems exist, relative humidity changes occur over a matter of hours (see Fig. 3-5).

Bulk polymer sensors, when exposed to rapidly changing humidity conditions, tend to become saturated with moisture.

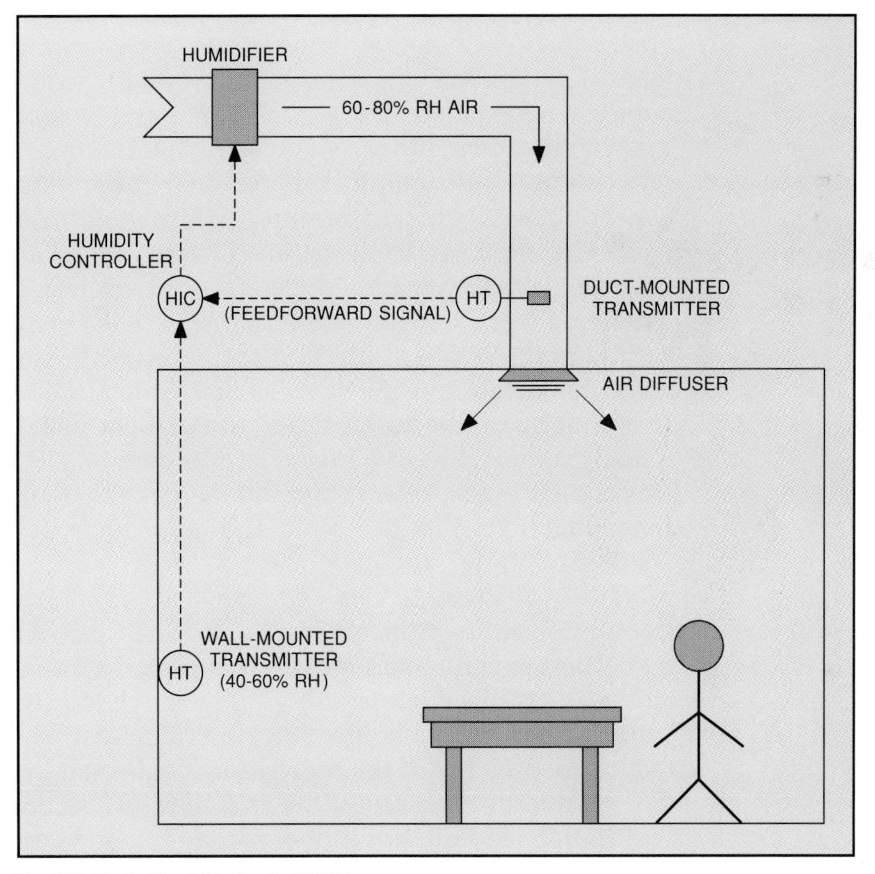

Fig. 3-5. Basic Humidity Control P&ID

Their recovery during a rapid downturn in humidity is a function of the time required for moisture absorbed by the sensor to evaporate. This, in turn, is a function of evaporation, a relatively slow process. Therefore, if humidity changes occur in the space at a rate faster than the evaporation process (and this is indeed possible), there will be an inherent lag in the humidity sensor response and its associated control loop.

In order to overcome this deficiency, a second humidity sensor should be installed in the supply air stream serving the monitored area. This duct-mounted sensor will detect increases in supply air humidity and provide a feedforward signal to the humidity controller. This will allow the controller to begin responding in advance of a change detected by the room sensor. The humidity controller can then react quickly to limit the amount of humidity entering the space and reduce, if not eliminate, space humidity swings.

While minor swings in space humidity do not, in most cases, create an environmental problem, these swings force the control system to cycle, causing additional wear on mechanical components (such as valves, actuators, and relays) and reduced system efficiency. Severe swings, depending upon the air temperature in the controlled area, may reach dewpoint. This condition results in condensation forming on the surface of equipment that can be disastrous in areas where corrosive materials, paper, powdered materials, or electronic equipment exist. In a worst-case situation, moisture can condense on the ceiling of the area and it will literally begin to rain indoors! This phenomenon is the same as that which causes moisture or frost to form on the surface of a cold glass or metal container quickly removed from a freezer or refrigerator. Moisture in the air contacting the outer surface of the container almost immediately reaches dewpoint and moisture from the air condenses on the surface.

Dehumdification of the air is accomplished mechanically in much the same manner. Moist air is passed through a cooling coil with a surface temperature low enough to cause the air contacting it to drop below dewpoint temperature. Moisture that condenses out of the air stream is collected in a pan below the coil and drained out of the airstream. By recirculating the air through the coil, additional moisture is removed. A heating coil downstream of the cooling coil reheats the air to a temperature suitable to maintain proper temperature in the controlled space.

As identified earlier in this section, transmitters such as the chilled mirror type directly measure dewpoint temperature, allowing the operator to assign alarm limits as ambient conditions approach dewpoint.

Temperature and relative humidity conditions can be plotted on a psychrometric chart to determine the dewpoint temperature (see Fig. 3-6). The set points for safety devices should be selected, based on the plotted conditions, and integrated into the control scheme to prevent excessive humidity conditions from occurring.

Good control system design practice dictates that safety devices be independent of loop control devices, since in some cases it may be a failure of one or more of the control devices that creates the fault condition in the first place. If it also served the role of safety device (for high-limit detection, for example) and failed, both control and protection capability would be lost. This design approach can, and should, be applied to any control system design.

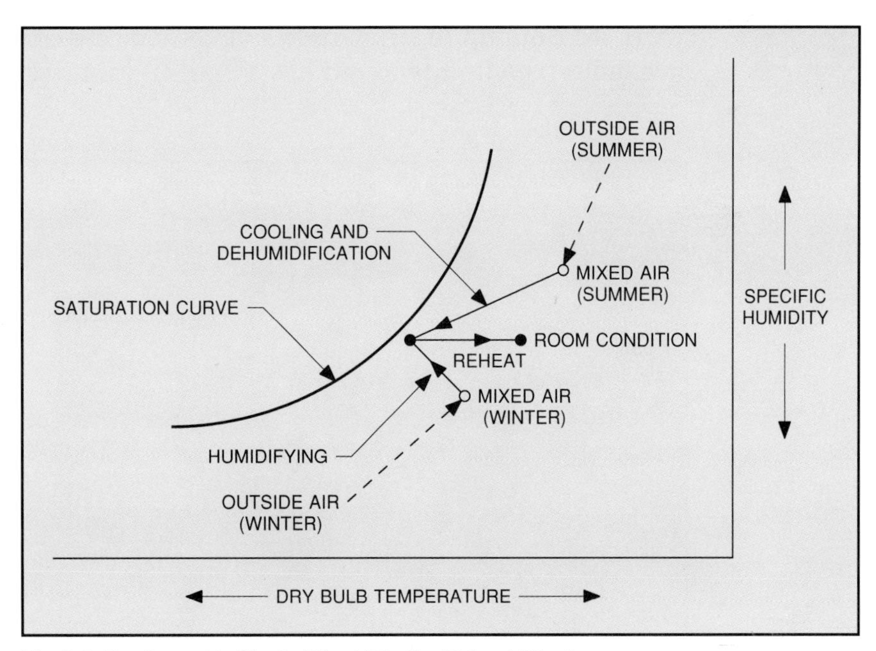

Fig. 3-6. Psychrometric Chart of Humidification/Dehumidification

Particle Counters

Measurement Process

Continuous and portable monitoring of airborne particulate is accomplished by frequent extraction and analysis of small sample quantities of air. Sampling rates will typically range from 0.1 to 1.0 cfm of air. Optical measurement is performed using a light dispersion technique to determine the quantity and relative size of particles present in the sample volume of air, similiar to turbidity measurement in liquids (refer to Unit 4).

A light beam passing through the air sample strikes airborne particles and is scattered. This scattered light is then optically collected by focal lenses and measured by a photo-sensitive diode. Dispersed light energy is then converted to electrical impulses that are proportional to the corresponding particle's diameter. These data pulses are counted, compared, and averaged electronically for a finite period of time to establish a time weighted average (TWA). This data is then recorded and displayed.

Portable, laser-based particle counters, such as that shown in Fig. 3-7, that can detect and count sub-micron size airborne particles from up to six sample sources simultaneously are available. The unit pictured has a built-in printer and optional

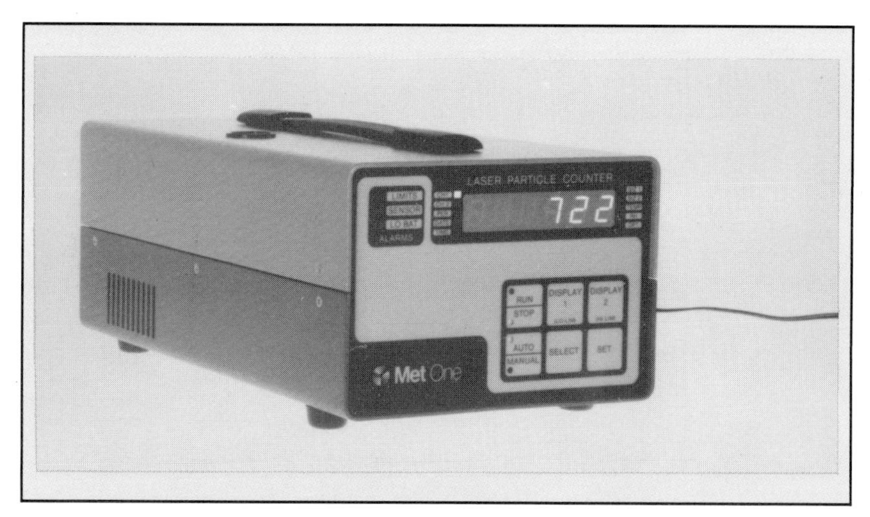

Fig. 3-7. Particle Counter (Courtesy of Met One)

attachments for measuring relative humidity, temperature, and air velocity in addition to particle count.

This type of monitoring system is particularly useful for establishing the performance level of clean room environments within a building, where particle concentration and relative size control are critical for an acceptable process environment. Control of particulate level is accomplished by a series of air filters and rapid air changes in the space. Filter systems are discussed in more detail later in this unit.

Dust Monitors

Dust monitors are used, as their name implies, to monitor levels of airborne dust, mists, fumes, and aerosols. These units are often used to evaluate hazardous (potentially explosive) dust concentrations in process areas. They are also helpful in determining sources of high dust levels, profiling dust patterns within a defined area, and evaluating the performance of ventilation systems and the relative effectiveness of air filtration systems.

High concentrations of suspended dust particles can create a very hazardous situation. The ignition level of dust from materials that, in a solid form, are generally not considered hazardous (such as flour or grain) have resulted in tremendous explosions and the loss of human life and property.

Dust monitoring systems can be interlocked to activate exhaust or makeup air fans in order to dilute the dust concentration before it reaches an explosive concentration. At the very least, the monitor can activate an alarm to notify equipment operators that they are approaching a hazardous condition and should take preventive action.

Bacteriological Air Sampling Systems

Bacteriological sampling systems aspirate variable volumes of air through a fine membrane. Any microbes in the air sample are impinged by air velocity onto an agar-surfaced contact plate (see Fig. 3-8). The sample plate is later removed and incubated. Incubation is accomplished by creating an ideal environment for bacterial growth (high humidity, warmth, food).

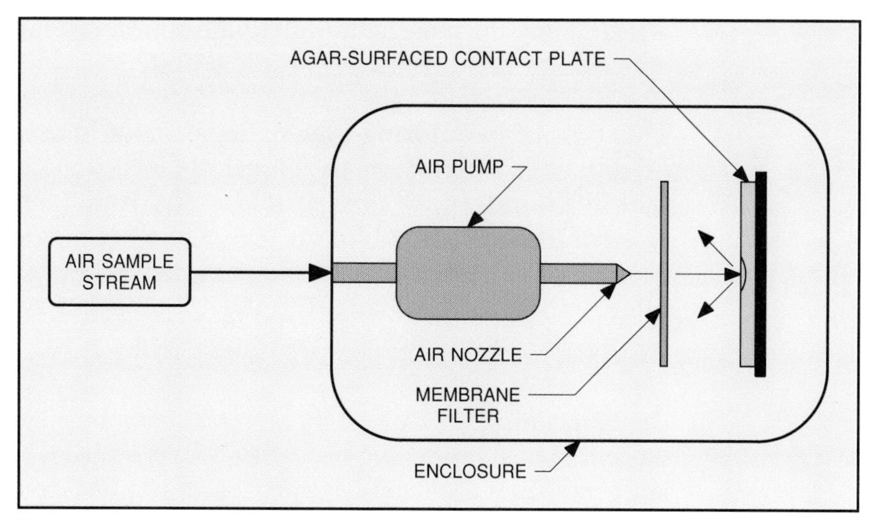

Fig. 3-8. Bacteriological Air Sampling System

After a few days, a microbe count is performed by a laboratory technician using a microscope to identify types of living bacteria present in the sample. This method of analysis is quantitative. It establishes the presence of different forms of airborne bacteria but does not fully determine the concentration. By controlling the incubation conditions, however, a relative concentration can be established by the number of organisms present after the predetermined incubation period. Biological air sampling systems that can sample and distinguish between both respirable and nonrespirable bacteria are also available.

Organic Vapor Analyzers

Organic vapor analyzers utilize a flame ionization detection (FID) process to measure total organic vapor (TOV) concentrations. This type of analyzer is often used to evaluate the presence of TOVs in and around hazardous landfills. These units are also used for detection of benzene leaks, stack monitoring, and arson investigations.

The FID process is also used for laboratory analysis of TOV concentrations.

Toxic Gas Survey Instruments

Toxic gas survey instruments employ cassette cards with a chemically impregnated tape to record toxic gas concentrations within an area. Concentration levels are determined by photoelectrically measuring the stain produced as contaminated (impregnated) tape is drawn through the instrument. This type of device is used as a quick and simple means to profile work areas in order to determine which personnel will require compliance monitoring (per OSHA regulations). They are also used to measure the effectiveness of ventilation and process controls.

Trace Gas Analyzers/Organic Vapor Meters

Trace gas analyzers or organic vapor meters utilize a photo ionization detection process to perform industrial hygiene surveys. They are also used to detect organic vapors from waste sites, residual solvent vapor from food or chemical processes, vinyl chloride within monomer plants, and low levels of benzene. These units are particularly useful in locating plant ''hotspots'' (areas of high vapor concentration), leak detection, and location.

Gas Detection Tube Sampling Systems

Gas detection tube sampling systems are used to determine concentrations of specific compounds in industrial hygiene surveys, ''tight building syndrome'' analyses, and other environmental studies. In this simple process, samples of gas or vapor are collected from suspect areas in sampling tubes. The samples are then analyzed by a mass spectrometer at a remote laboratory.

Combustion Gas Monitors

Combustible gas monitors are used to monitor flammable gas levels in terms of their LEL (lower explosive limit) (see Fig. 3-9). These units are installed in areas in which the potential exists for gas or vapor concentrations to exceed their LEL. Examples would include flammable liquid or gas storage facilities (where accidental spills or leaks could occur) and battery rooms (where battery charging produces hydrogen gas).

Detection units are typically interlocked with mechanical exhaust or makeup air systems (1) to provide emergency ventilation (or dilution) at concentrations approaching the LEL and (2) to provide remote alarm annunciation and equipment

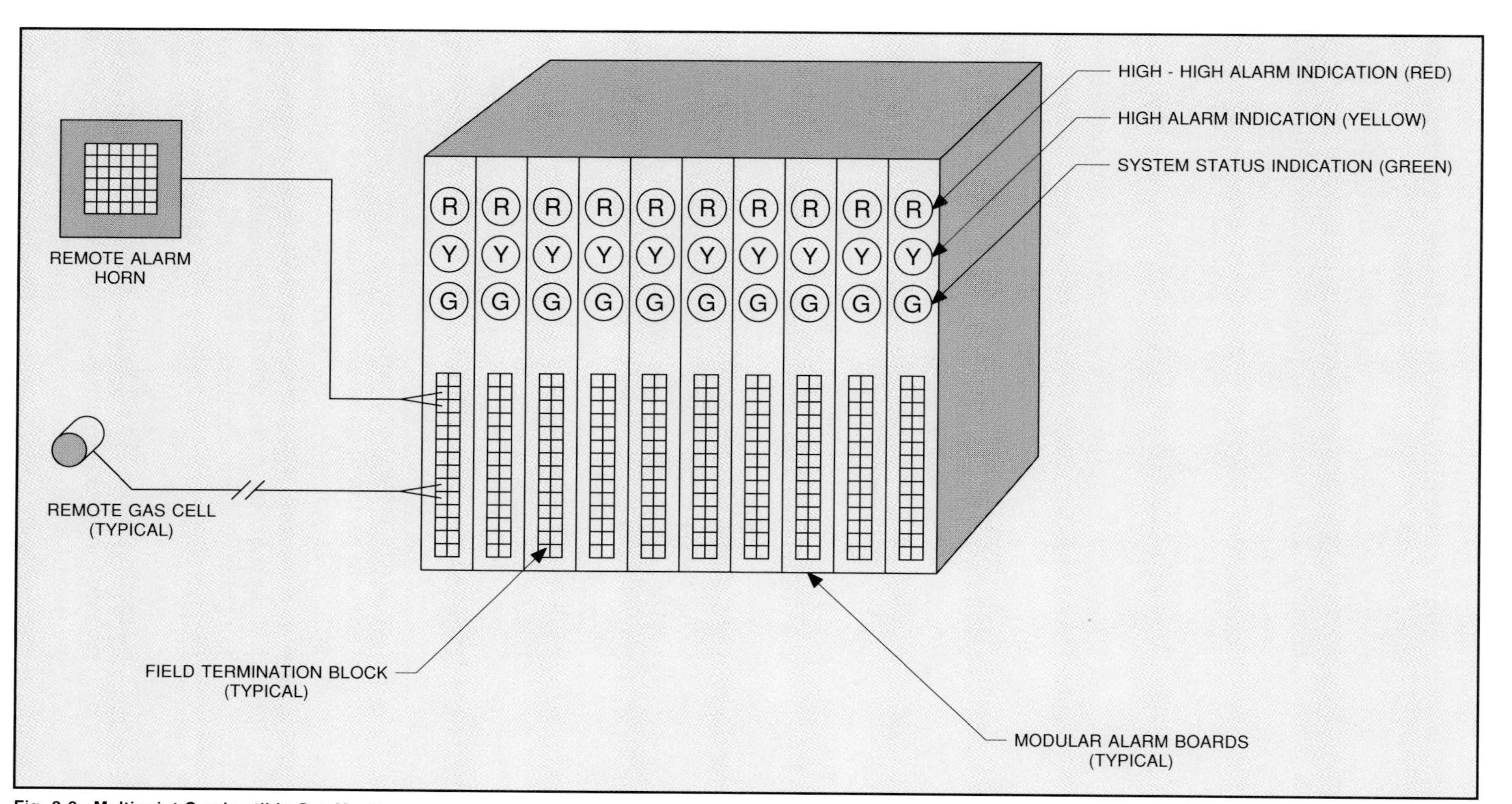

Fig. 3-9. Multipoint Combustible Gas Monitor

shutdown to prevent a potentially explosive condition when the gas concentration reaches or exceeds its LEL.

Gas Leak Detectors

Simple gas leak detectors use a thermal conductivity sensor to detect small leaks of any gas mixture that has a thermal conductivity greater than that of air (reference gas).

A typical application for this type of detection system would be a tank farm where a single known gas is stored and potentially hazardous leaks can occur. The device does not distinguish one type of gas from another. However, in this case it does not matter: the gas type is known. All the owner needs to know is whether or not a leak exists.

Auxilliary Devices

Air Sampling Pumps

Air sampling pumps allow low flow rate sampling of airborne particles and toxic gases using sorbent tubes, filters, impingers, or sample bags.

Stack Sampling Probes

Stack sampling probes allow portable ambient air monitoring instruments to be used as continuous source gas, or stack, analyzers (see Figs. 3-10 and 3-11). These probes are capable of monitoring stack gas at temperatures up to 750°F and opacities up to 100%.

In this process, stack gas is continuously extracted, drawn through a sonic orifice, and then diluted with dry instrument air to a ratio of approximately 200 to 1. Diluted sampling eliminates problems associated with moisture in the sample gas that condenses when it comes into contact with the much cooler measuring equipment. For the same reason, this also eliminates the need to heat-trace sampling lines.

Sampling is performed under positive pressure, eliminating potential errors that can occur with negatively pressurized sampling due to line leakage (air infiltration). Samples are

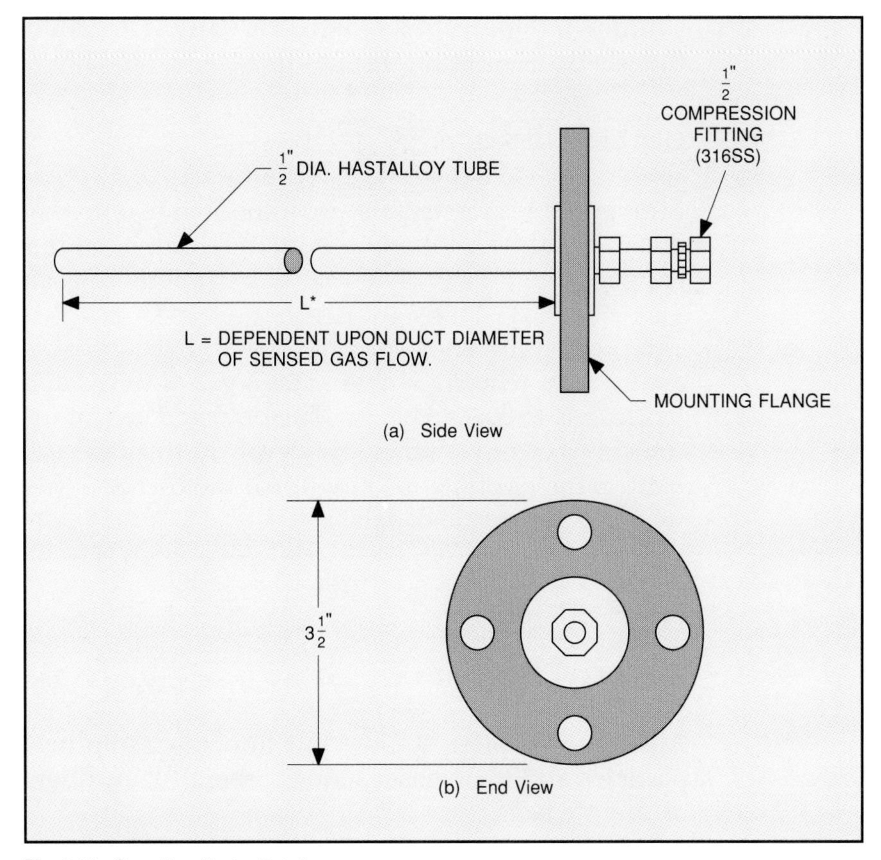

Fig. 3-10. Sampling Probe Detail

measured wet, which is the EPA's preferred method of analysis.

Gas Chromatography (GC)

Three types of detectors are commonly used for gas chromatography: thermal conductivity, infrared absorption, and mass spectroscopy.

Thermal Conductivity Detector (TCD)

The thermal conductivity detector is one of the most widely used because of its construction, ruggedness, versatility, linearity over a wide range, and relatively low cost.

A TCD typically consists of two metallic blocks, each containing a tubular cavity for the flow of gas. A heated

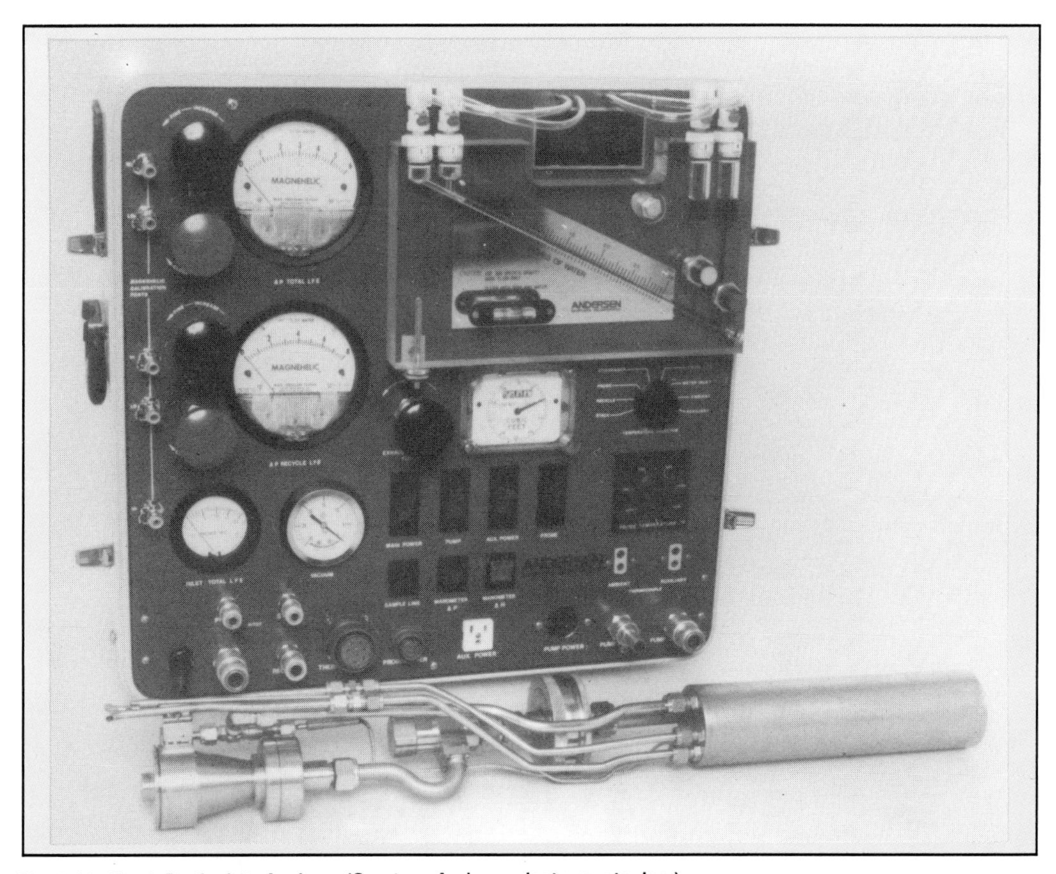

Fig. 3-11. Stack Particulate Analyzer (Courtesy Andersen Instruments, Inc.)

resistance element or thermistor located within the cavity dissipates heat to the block at a rate dependent upon the thermal conductivity characteristics of the flowing gas. A ''carrier'' or reference gas flows through one block and sample gas through the other. Heated elements in each block are electrically connected to form a Wheatstone bridge (see Fig. 3-12).

When the same gas is passed through both cavities simultaneously, the bridge network is balanced by manually adjusting the balancing potentiometers. When balanced, electrical output of the circuit is zero volts. When the thermal conductivity of one of the gas streams changes, the temperature (and resistance) of the detector element (constituting one leg of

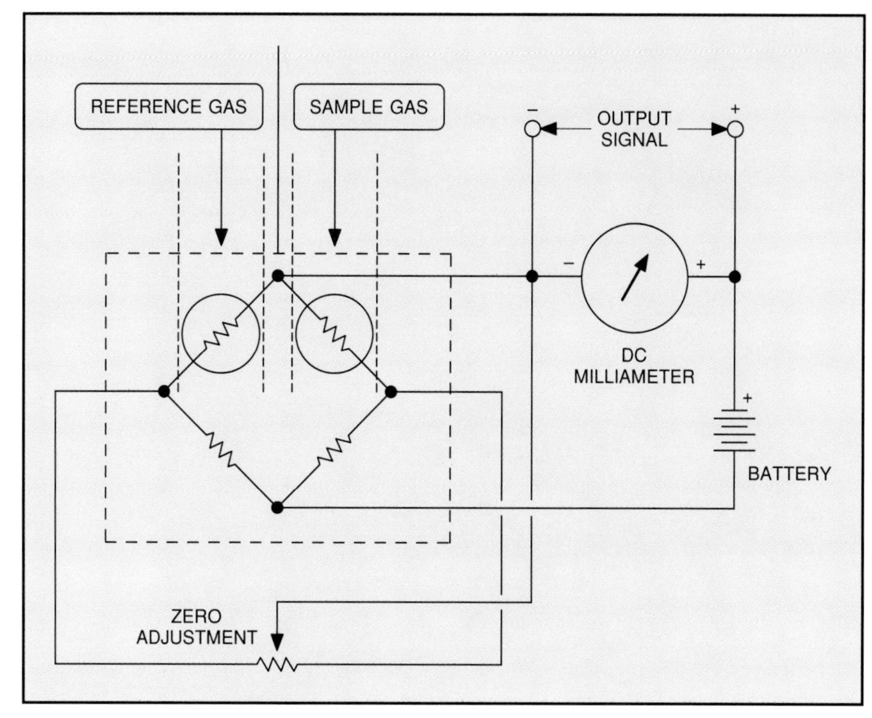

Fig. 3-12. Thermal Conductivity Analyzer

the bridge) changes proportionally, causing an imbalance of the bridge and an incremental increase in the circuit's output signal.

Infrared Absorption

Infrared absorption (IA) is essentially a simplified version of spectrophotometry. An elaborate spectrophotometer examines absorption of gas components over the entire light spectrum. IA, by comparison, uses a relatively simple filter photometer for analysis.

Infrared absorption analysis is an effective method for identifying components within a gas stream when the component being analyzed has an infrared absorption band (frequency) in a region where other present gas components do not.

Mass Spectrometry

Mass spectrometry (MS) is a very effective technique for analysis of unknown compounds. A very small sample of an

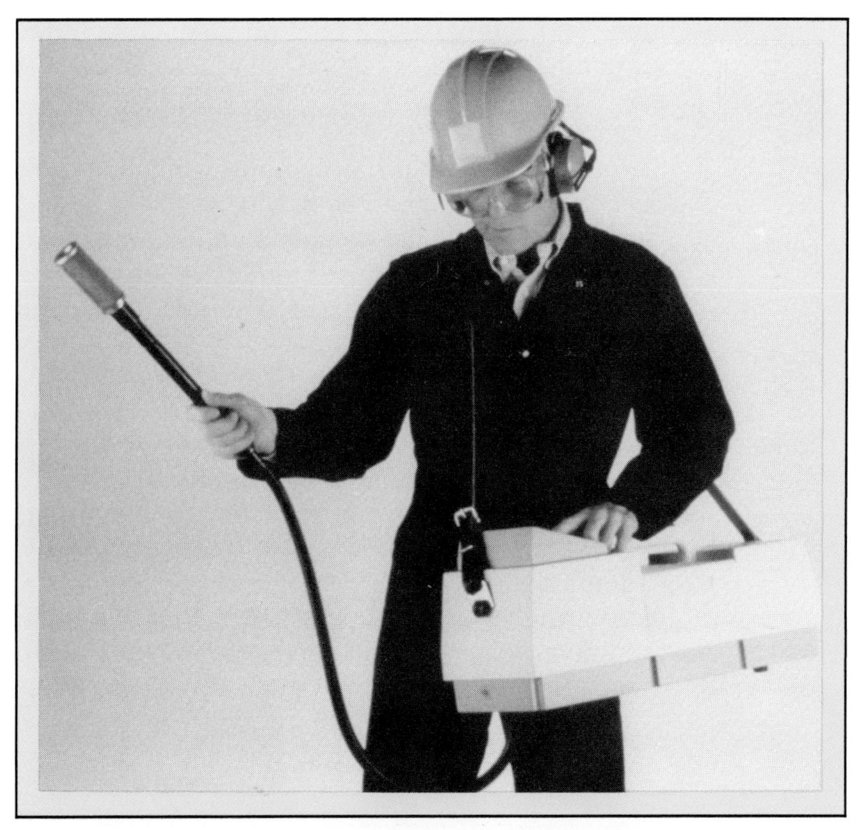

Fig. 3-13. Portable Infrared Gas Analyzer (Courtesy The Foxboro Company)

unknown material is ionized and passed through a magnetic field. Molecular mass of the sample fragments can then be counted and measured. This fragment pattern is then recorded and matched with a known structure pattern to provide fairly conclusive identification of the compound.

Since mass spectroscopy relies on differences in the mass of gas constituents for identification, isomers (which have the same mass) cannot be identified by this type of analysis.

Air Quality Sensors

Continuous measurement of air quality in commercial facilities can be an expensive investment if monitoring of numerous zones is required. Several commercial control manufacturers have developed and are refining relatively inexpensive air quality sensors that are used to measure the percentage of oxygen or carbon dioxide present. However, these parameters alone (oxygen or CO_2) do not provide a true indication of the air quality but rather ''infer'' what the condition of the air might be.

The cost of analyzers that will measure all of the components needed to determine total air quality (oxygen, carbon dioxide, carbon monoxide, gas composition, particulate, and bacterial levels) would be very cost-prohibitive for commercial applications. Their use has been limited to laboratory, process, or manufacturing environments.

Multipoint ambient air monitors that will continuously monitor low (ppm) to high (percentage) gas concentrations in up to 24 zones within a building are presently being used for employee exposure monitoring, leak detection, and ventilation control.

Examples of the type of facilities in which these multipoint monitoring systems are currently being used include the following.

Fig. 3-14. Multipoint Monitor (Courtesy The Foxboro Company)

For OSHA Compliance

- Polymer Manufacturing Facilities
- Hospital Operating Rooms
- Fumigation Facilities
- Degreasing Areas
- Magnetic Tape Manufacturing Facilities

For Leak Detection

- Refineries
- Tank Car Loading/Unloading Areas
- Pesticide Manufacturing Facilities
- Rocket Fueling Areas
- Solvent Transfer Lines
- Refrigeration Units

For Perimeter Monitoring

- Hazardous Waste Sites
- Nuclear Power Plant Control Rooms

(Source: The Foxboro Company)

These monitor systems use a single-beam, microprocessor-controlled infrared spectrometer for gas detection and analysis. Up to five individual compounds in the air sample can be monitored simultaneously. Samples from zones within 50 feet of the monitor can be analyzed in less than 30 seconds. Samples from locations 1000 feet away require approximately 60 to 90 seconds for analysis.

Analytical results are presented as 8-hour time weighted averages (TWA) for each component at each sample location. The report produced also identifies the highest concentration that occurred during each shift at each zone for each component, and the time at which it occurred. Carrier gases are not required. Maintenance is limited to periodic replacement of a particulate filter at the end of the sample lines.

Environmental Corrosion Potential

Corrosion is a serious concern in any buildings used for storage and handling of solvents, acids, caustics, and some strong cleaning solutions containing the aformentioned characteristics.

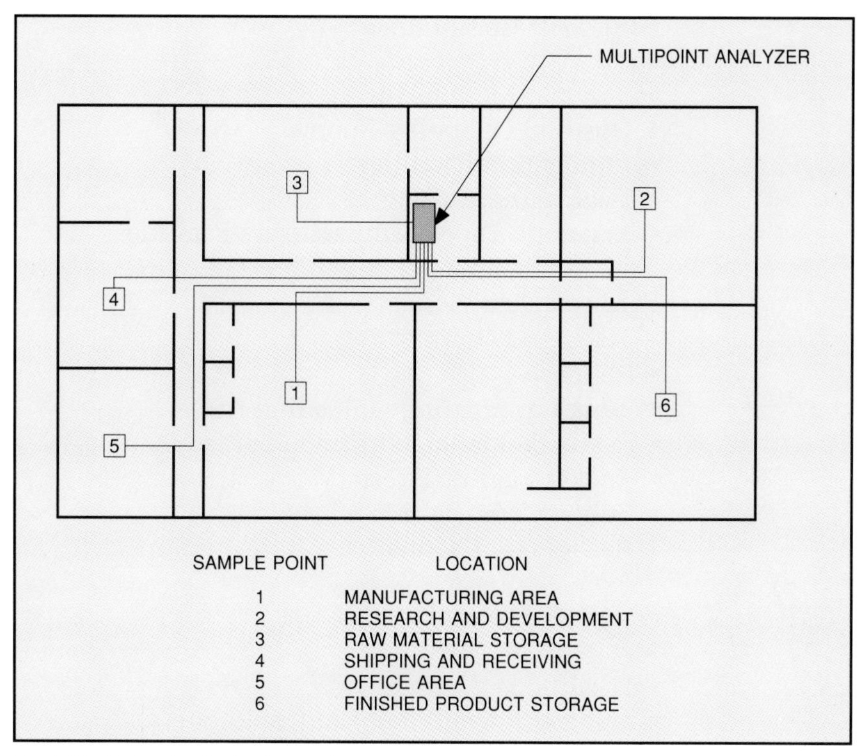

Fig. 3-15. Multipoint Air Quality Analyzer Floor Plan

Replacement of damaged equipment can be costly and time consuming. Severe corrosion left unchecked can eventually weaken and damage the structural integrity of the building steel.

How is the corrosion severity in a building area determined? Extensive research conducted by Battelle Laboratories and Bell Telephone Company to answer this question has led to the development of a relatively simple, inexpensive method of surface film analysis referred to as cathodic/electrolytic reduction.

In this analytical process, a thin metallic or foil film-covered "coupon" (typically copper or silver) is placed in an electrolyte solution. A constant current is introduced between the coupon, which acts as a cathode, and a platinum anode reference. This apparatus is then placed in the area to be tested.

Thickness of the resulting corrosive film (measured in angstroms) forms at a rate representative of the corrosion potential of the surrounding air. Corrosion severity is a

	Class G1	Class G2	Class G3	Class GX
Copper coupon film thickness (angstroms)	0–299	300–999	1000–1999	>1999
Gas concentrations (parts/billion)				
Hydrogen sulfide	<3	<10	<50	>50
Sulfur dioxide	<10	<100	<300	>300
Chlorine	<1	<2	<10	>10
Nitric oxide	<50	<125	<1250	>1250
Hydrogen fluoride	<1	<2	<10	>10
Ammonia	<500	<10000	<25000	>25000
Ozone	<2	<25	<100	>100

function of the corrosive gas or vapor concentrations present in the air.

The Instrument Society of America (ISA) developed the standard ISA-S71.04-1985, Environmental Conditions for Process Measurement and Control Systems: Airborne Contaminents, which serves as a means to classify an environment by its overall corrosion potential and severity level (refer to Table 3-3). The classifications in the table identify ranges of corrosive gas concentrations that increase the corrosive film thickness of a copper test coupon within a defined time period.

Some companies offer an environmental service in which the test apparatus described above is set up at the test site. After a designated period of exposure, the coupon is retrieved from the test site and sent to a laboratory for analysis. A report on the findings is sent to the facility owner's representative.

3-4. Air Quality Control Methods

Air Filtration

As summarized earlier in this unit, airborne particles contribute significantly to poor air quality. This can be supported by anyone who frequently suffers from allergies.

Air filtration systems have been designed that can significantly reduce the levels of respirable particulates entering a conditioned space.

The chart in Fig. 3-16 identifies the relative size of suspended airborne particulates. A large percentage of the airborne bacteria, mold, smoke, and dust fall within a size that can cause lung damage (respirable range). Not identified on this chart are microscopic asbestos fibers, which also fall within the respirable range. Asbestos fiber has been linked to several forms of cancer.

Air filtration systems are identified chiefly by their capture efficiency rating for various particle sizes. When an application requires removal of particles from an air stream that vary greatly in size (diameter), a series of filter media may be required, each filter capable of removing smaller diameter particles than the previous filter. This prevents total blockage of air or gas flow through the filter media, a condition generally found undesirable. The ''coarse'' filters capture the larger diameter particles while allowing smaller particles to continue on, where they are then captured by a higher efficiency ''fine'' filter medium.

As can be seen in the graph in Fig. 3-17, the efficiency of the filter media diminishes as the particle size becomes smaller.

Air filter designs and their relative efficiency for removal of submicron, lung-damaging particulates are found in Table 3-4.

Some types of exhausted particulate cannot be removed with standard filter media. An example are dusts created by the grinding of magnesium or aluminum. These dusts, in the right concentration, can be flammable and potentially explosive. Filter material used in a dry air filtering process is often flammable. Therefore, these types of particulate must be removed from the air stream through a wet filtering system. Such a system typically recirculates scrubber water, which is sprayed across the air path. Particulates are knocked out of the air stream by atomized molecules of water and carried with the water to a settling tank that allows the particulate to settle to the bottom, forming a sludge material. The sludge is periodically removed and disposed of as waste.

The flammability and LEL (lower explosive limit) of any particulate material must be reviewed to determine what is required in order to safely treat it.

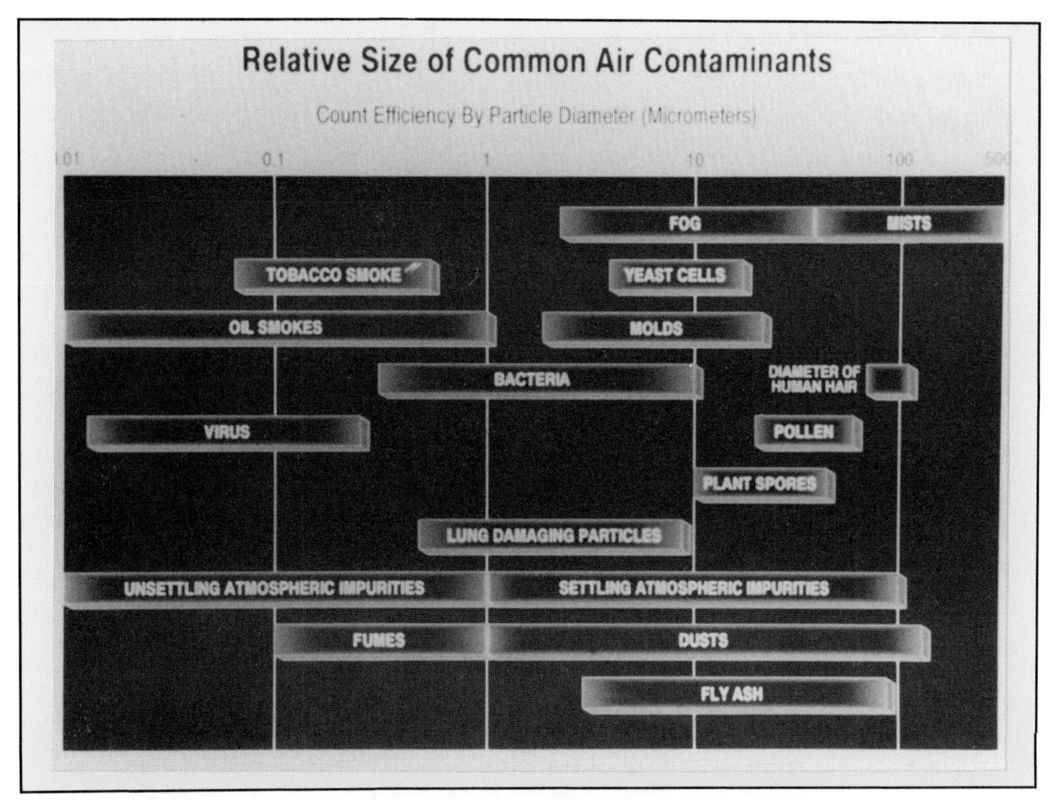

Fig. 3-16. Relative Sizes of Suspended Airborne Particles (Courtesy Cambridge Air Filter)

Type	Typical Efficiency
Disposable panel filters	<20%
Washable panel filters	<20%
Media pad filters	<20%
Roll filters	<20%
Pleated surface filters	25–35%
Bag filters	<20%
HEPA filters	99.9%
Electrostatic filters	90.0%

Table 3-4. Air Filter Efficiencies (Source: Cambridge Air Filter)

Monitoring Filter Status

If the filter medium is allowed to collect enough particulate to greatly restrict gas or air flow, a large pressure differential will develop across the filter due to the significant pressure drop. If the system is designed to force air through the filter (blow-through system), pressure will develop upstream of the filter, and the downstream side will probably approach atmospheric pressure.

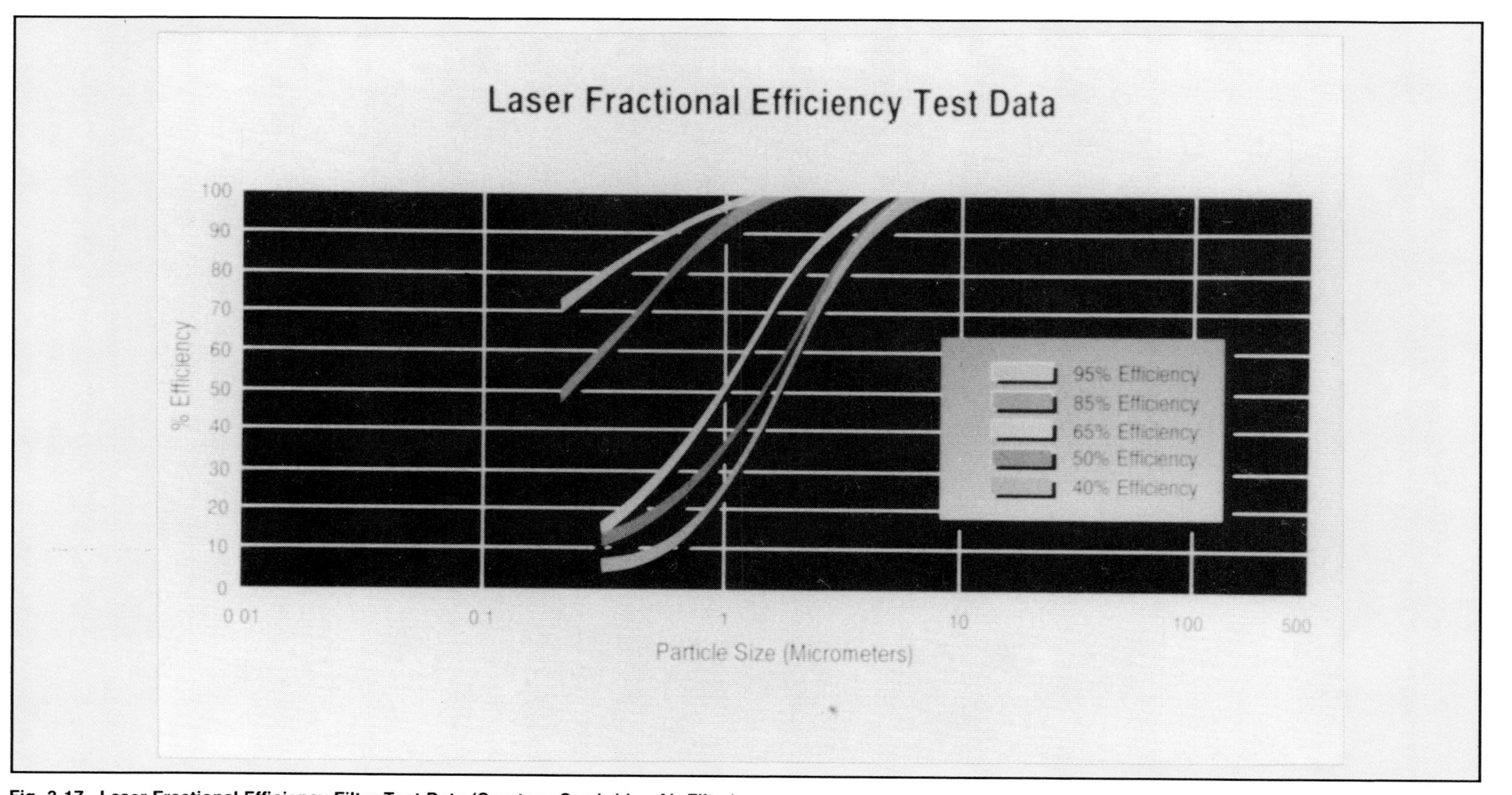

Fig. 3-17. Laser Fractional Efficiency Filter Test Data (Courtesy Cambridge Air Filter)

If, on the other hand, the system is designed to draw air through the filter (draw-through system), the downstream side will develop a negative pressure as the upstream side approaches atmospheric.

In either case, if the pressure difference across the filter becomes great enough (dependent upon construction of the filter section), the filter may collapse, resulting in major contamination of the downstream ductwork and supplied areas or systems.

Frequent inspection of the filter medium can prevent this. By installing a differential pressure gage piped to either side of the filter medium, changes in pressure drop can be easily noted. As the pressure difference reaches the maximum acceptable limit, the medium is replaced. Several methods of monitoring pressure differential (in inches of water) can be employed. For simple local indication, a differential pressure gage that has a range low enough to measure changes as small as one-half inch of water can be used.

For remote monitoring, a draft range differential pressure transmitter can be piped independently or in parallel with the manometer to provide a 4-20 mA DC signal to a remote display, controller, data collection system, or DCS (distributed control system).

High alarm limits can be assigned to the remote signal to notify the system operator when the filter medium is dirty enough to be replaced and when the differential pressure has reached a critical point at which the filter may collapse (high-high alarm). The rate of change can also be measured and updated on the DCS, allowing the rate of filter changes to be predicted for future preventive maintenance requirements.

Humidity

Humidity extremes can be avoided by (1) providing suitable humidity sensing and control that limit the amount of moist air allowed to enter the space or (2) by using a dehumidification process to condense moisture out of the air.

Limiting the quantity of moist air introduced into a building, while at the same time providing sufficient ventilation, requires direct measurement and control of the supplied air's relative humidity. If a sufficient quantity of recirculated (return) air

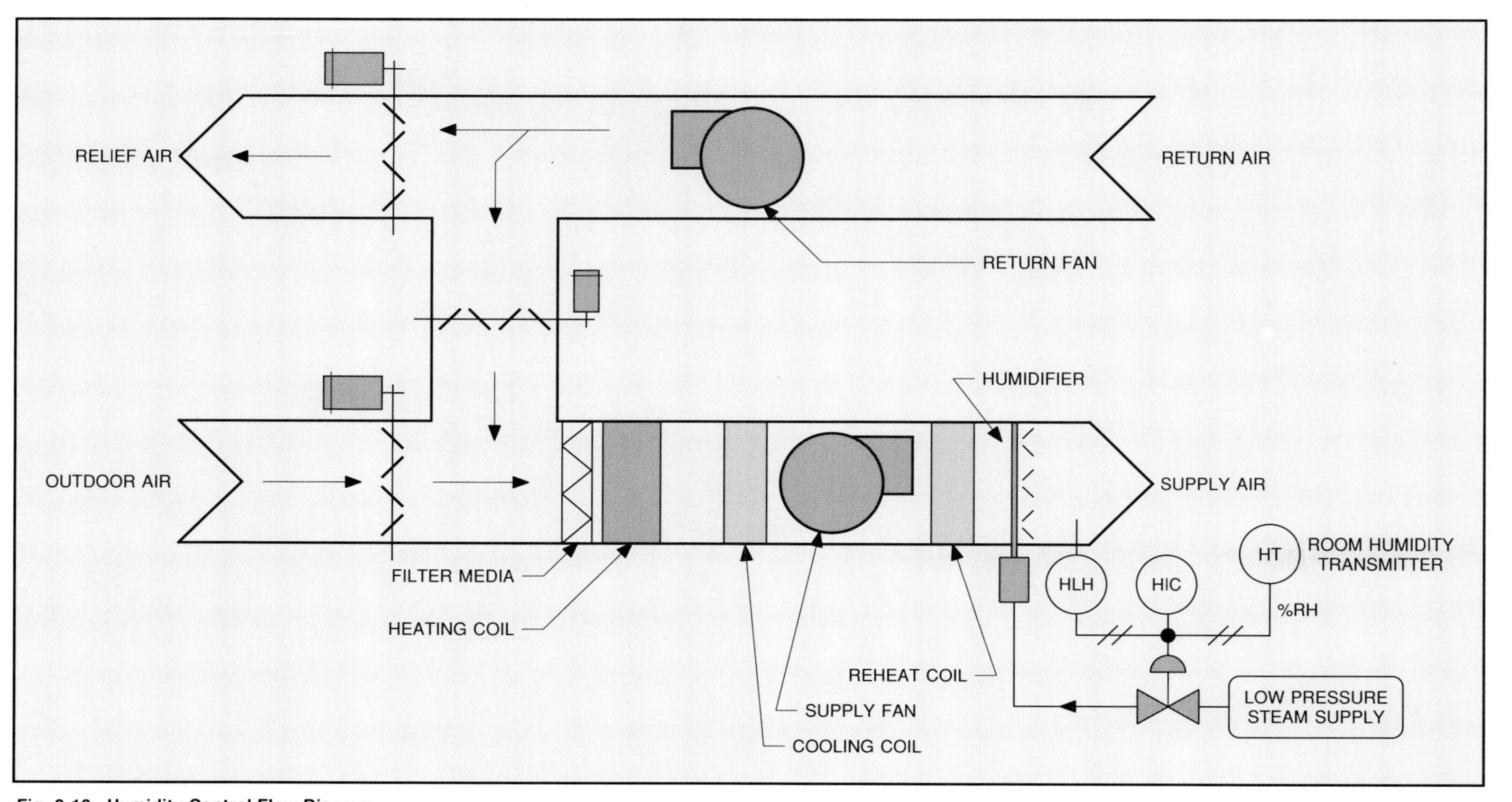

Fig. 3-18. Humidity Control Flow Diagram

from the controlled space is mixed with incoming outdoor air so that supply air humidity is maintained within the optimum 40 to 60% range, then no dehumidification should be required.

If, for example, a large percentage of fresh air is required for ventilation and the outdoor air relative humidity is very high (say, 100%), some form of mechanical dehumidification will likely be required to prevent the inside humidity from exceeding 60%. The two methods generally used to accomplish this are dessicant and refrigerant systems.

Dessicant (drying) systems utilize a dessicant material that absorbs moisture out of the air as it passes through the porous material (see Fig. 3-19). Salts are typically used as a dessicant material. This type of drying system is very effective when very low humidity conditions are required, such as in controlled laboratory environments.

However, this type of system can also be very costly when large volumes of conditioned air are required. It is not unusual for a system conditioning 20,000 cfm of outdoor air to 15% RH to have an installed cost approaching $250,000 (based on typical 1990 material and labor costs).

Contributing to the high equipment cost is the need for two dessicant chambers. Dessicant is capable of absorbing a finite amount of moisture. It must then be dried out (regenerated) in order to again absorb moisture. In order to accomplish this, and at the same time maintain a continuous dehumidification process, a second dessicant chamber must be used during regeneration of the primary chamber.

Regeneration is accomplished by heating the dessicant to boil off the absorbed moisture. The resulting water vapor is then exhausted into the atmosphere.

In addition, a relatively high pressure drop is created by the dense dessicant chamber, resulting in a greater fan horsepower requirement than that of mechanical/refrigerant drying systems.

To fully understand the psychrometrics of dehumidification requires a familiarity with the psychrometric chart developed by the late Dr. Willis Carrier (see Fig. 3-20). While developing the process of mechanical refrigeration, Dr. Carrier was successful in graphically representing the physical relationship between air's temperature, moisture content, relative humidity,

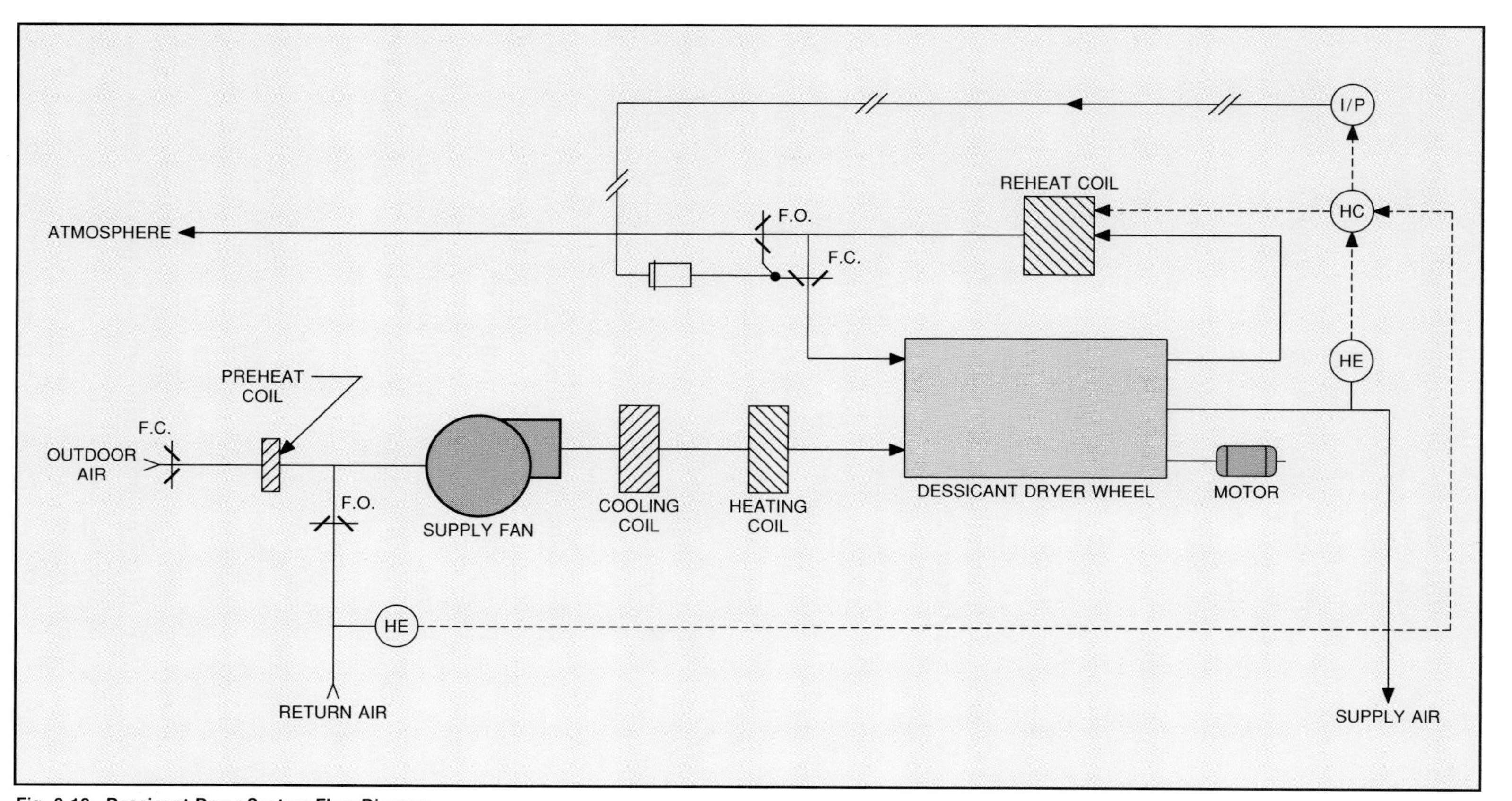

Fig. 3-19. Dessicant Dryer System Flow Diagram

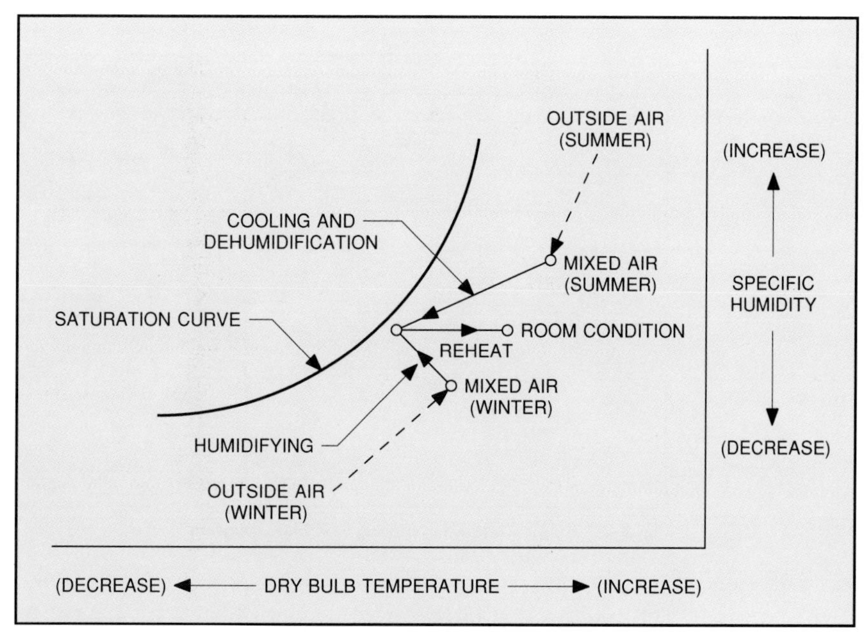

Figure 3-20. Psychrometric Chart of Humidification/Dehumidification

and dewpoint (temperature at which moisture condenses out of the air).

Air conditioning system engineers must specify systems that can reliably and efficiently deliver a sufficient quantity of fresh air to maintain a healthy and comfortable environment within occupied spaces. To accomplish this feat, these systems must also be capable of providing sufficient air velocity at air diffusers within the space to allow adequate distribution and mixing of the fresh air once it reaches the conditioned space (see Fig. 3-21).

Chemical Contaminants

In an industrial environment, multipoint ambient air monitors (see Fig. 3-22) employing infrared spectrometry can be used to detect areas in which exposure levels of potentially harmful chemical compounds are approaching OSHA limits. Refer to Appendix D for a list of potentially hazardous compounds and the current OSHA exposure levels.

The monitoring system, in turn, actuates a change in dilution rate of the air in that area or zone of the building by increasing the exhaust rate or makeup air rate. The gas concentration is reduced as the area is purged with fresh air.

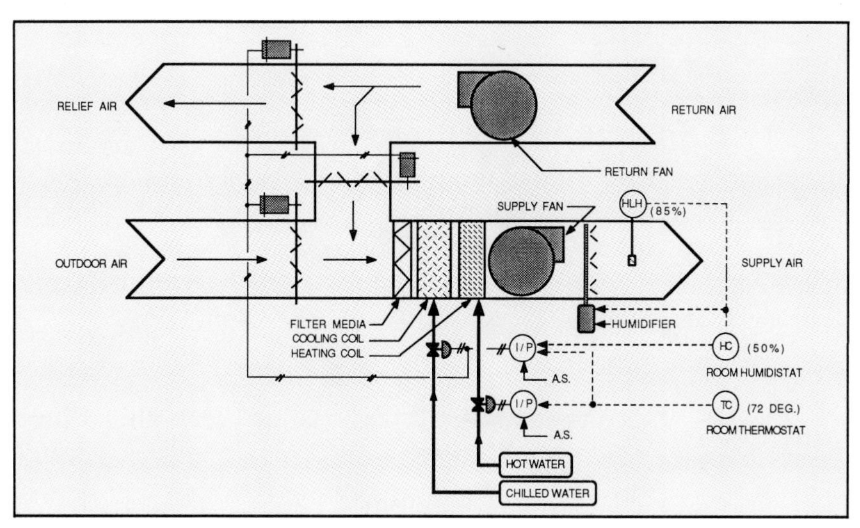

Fig. 3-21. Single-Zone Air-Handling Unit Flow Diagram

The same holds true for areas where potentially flammable or explosive gas vapors can collect. By increasing the dilution rate, gas concentrations are reduced and do not reach their LEL (lower explosive limit). For heavy gases (those denser than air),

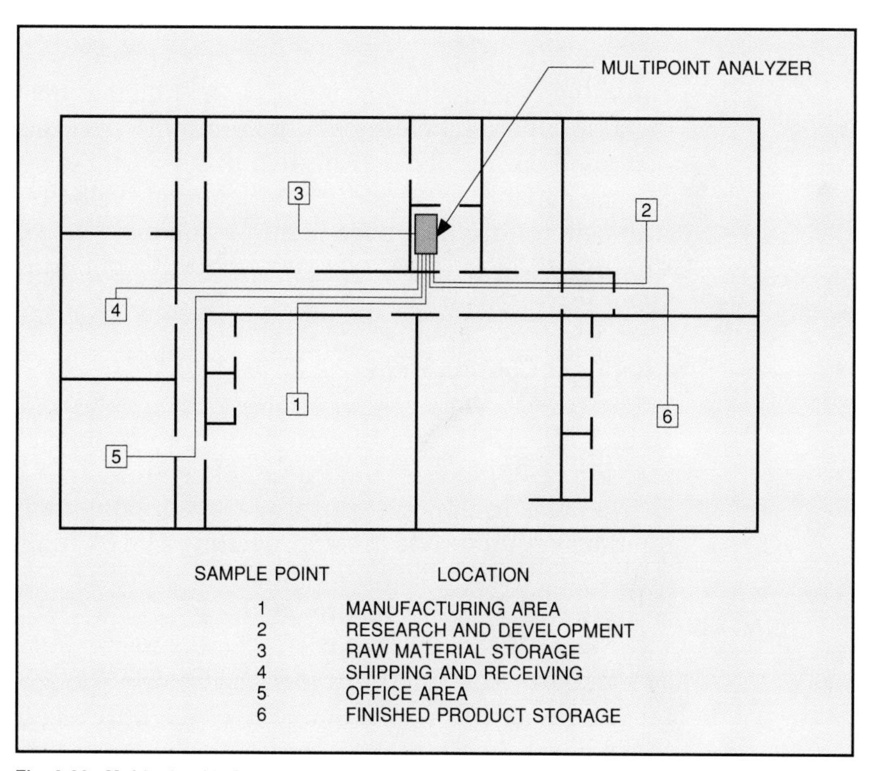

Fig. 3-22. Multipoint Air Quality Analyzer Floor Plan

the pattern of air flow in the area must also be examined to ensure that pockets of gas do not collect and go undetected by the air monitoring system.

Continuous monitoring points or zones are effective in detecting the presence of a contaminant. The leak (source) can then be pinpointed, using a portable analyzer, and corrected.

Corrosive Environments

Reducing the effects of a corrosive environment can be accomplished by (1) increasing the ventilation rate, (2) increasing dilution of the gas or vapor concentration using outdoor air, or (3) reducing or eliminating the source of the gas or vapors (source reduction).

3-5. Summary

Poor indoor air quality often results in occupant discomfort, potential illnesses, reduced worker productivity, and potential liability associated with related health effects. The best method of control, in most cases, is to increase the dilution rate by introducing more fresh air from makeup air units and exhausting fumes from the building. A compromise must be established between energy consumption and the risks associated with the "sick building syndrome."

References

1. Howell, Ronald H., and Sauer, Jr., Harry J., *Environmental Control Principles*, American Society of Heating, Refrigeration and Air-Conditioning Engineers, Inc. (1985).
2. ASHRAE Standard 62-1989, "Ventilation for Acceptable Indoor Air Quality," American Society of Heating, Refrigeration and Air-Conditioning Engineers, Inc. (1989).
3. Brownawell, Mark, "An RH Sensor Review, with HVAC Considerations," *Sensors*, March, 1989.
4. Duffy, Gordon, "A Prescription for Healthy Buildings: Ventilation Efficiency," *Engineered Systems*, September/October, 1990.
5. Meckler, Milton, "Evaluating Indoor Air Quality," *Heating/Piping/Air Conditioning*, September, 1990.
6. *1986 Industrial Ventilation—A Manual of Recommended Practice*, American Conference of Governmental Industrial Hygienists (ACGIH) (1986).
7. ANSI/ASHRAE Standard 55-1981, "Environmental Conditions for Human Occupancy," American Society of Heating, Refrigeration, and Air-Conditioning Engineers (1981).
8. ANSI/ISA-S71.04-1985, Environmental Conditions for Process Measurement and Control Systems: Airborne Contaminants, Instrument Society of America (1985).

Exercises:

3-1. *List three common contributors to poor indoor air quality.*

3-2. *What type of analyzers are effective for measurement of particulate quantity and size in an industrial environment?*

3-3. *Sketch and describe the basic components of an infrared absorption-type gas analyzer.*

3-4. *Explain the purpose of a combustible gas monitor.*

3-5. *Identify the advantages and disadvantages of current technology air quality sensors used for environmental control.*

3-6. *Identify two types of air filtration systems that efficiently remove airborne particulates and explain their advantages and disadvantages.*

3-7. *Design and sketch control systems that could be added to the air-handling system below to improve indoor air quality.*

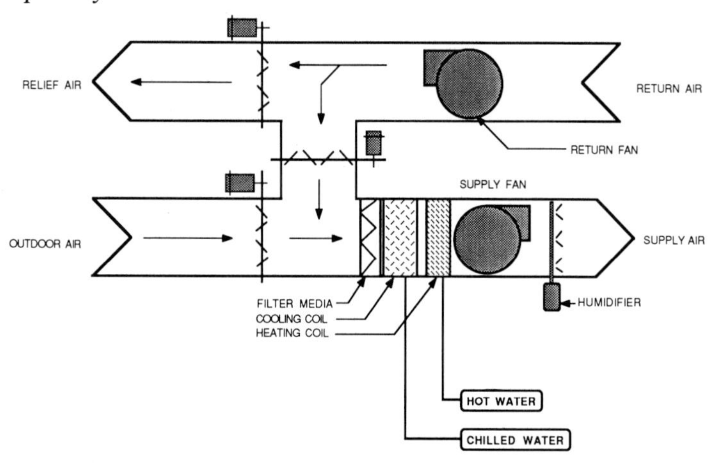

3-8. *Describe the effect the systems used above will have on the air quality and HVAC system performance.*

3-9. *Design a humidity control system that will maintain a constant relative humidity range of 40–50%. Include controls necessary to prevent the humidity from exceeding 70%.*

3-10. *Describe a control system that could maintain relative humidity at 15%. In what type of application or facility might you find this type of control requirement?*

Unit 4:
Waste Stream Pretreatment

UNIT 4

Waste Stream Pretreatment

This unit discusses the pretreatment of wastewater in order to make it suitable for release into the environment. Particular emphasis is placed on industrial wastewater effluent. Some of the processes described in this unit also apply to commercial facilities. Local regulations should be reviewed to determine what control measures are necessary for a specific facility and its effluent.

Learning Objectives — After you have finished this unit, you should:

A. Be familiar with instrumentation typically used to measure and control wastewater emissions.

B. Know what type of instream analysis technique is required for a typical monitoring or control application.

C. Understand pH measurement and control of wastewater.

D. Be familiar with automatic sampling systems, how they are applied, and their advantages over manual sampling methods.

E. Know the basic methods of open channel flow measurement.

F. Be familiar with fundamental components of wastewater treatment processes.

4-1. State and Federal Regulations

State and local regulations applicable to wastewater effluent vary greatly by geographical area. Therefore, EPA regional representatives should be contacted for information on regulatory requirements that apply to the area in which one will be working. Addresses and phone numbers of the EPA representatives, by region, are found in Appendix C of this text. Local regulatory environmental agencies can be contacted by consulting the phone book.

It is important to remember that local regulations are often more stringent than EPA requirements. Therefore, before proceeding with the selection of instrumentation for measurement and control of wastewater, be sure you fully understand what level of control or measurement accuracy will be required to satisfy not only current regulations but any proposed regulations as well.

This is not to say that you should go out and buy instrumentation or systems with the highest accuracy or most precise control. They may not be necessary to meet current or proposed regulations, and a very significant cost may be associated with such a decision, which you will have to justify later on.

However, the intent of the regulations is to protect people and the environment from possible harm. This moral responsibility is yours as well as the regulatory agency's. Work together to provide a safe, reliable system and environment.

4-2. Instream Analyzers

Sampling Methodology

Accurate and precise wastewater analysis depends upon instrument accuracy, how representative the test samples are of the true effluent condition, and reliability of the overall analysis process. For example, samples obtained and analyzed at a frequency of once every four hours from a waste stream having very erratic characteristics (such as large variations in flow rate, chemical makeup, or concentration over a matter of minutes) will *not* be very representative of the average effluent conditions. However, on waste streams with very stable characterisitics, this sample rate may be satisfactory.

Preliminary Testing

Continuous sampling and analysis processes should not be designed and installed until the waste stream conditions have been properly investigated.

Sampling rates should be determined in advance through preliminary ''manual'' sampling in order to determine frequency and magnitude of variations in effluent makeup.

In order for preliminary sampling to be effective, the sampling technician should first obtain any available information about the source and characteristics of the process or processes that create the effluent. This data can usually be found within application data for any existing discharge permits. The best source of such information is typically a facility engineer, plant engineer, or someone who acts within that capacity. At large process facilities, there may be environmental engineers or technicians on staff who are specifically responsible for maintaining this data.

In some cases, the facility's engineers or technicians, for whatever reason, may not know and cannot predict the range of effluent characteristics that exist or are anticipated. Environmental permits may not already exist, or processes within the plant may have changed significantly, which would make any available information invalid.

Information requested about the processes that create waste effluent should include a list of chemical constituents that make up the waste stream, their concentrations, volumetric flow rate, and temperature. At some plants, waste production may continue 24 hours per day. At others, effluent throughput may vary dramatically with time, depending upon production schedules and temporary storage capacity (effluent may be accumulated in an on-site storage tank(s) for pretreatment before its release).

When little or no preliminary information is available, an initial test frequency of several samples per hour (say, at 15 minute intervals) should suffice until the stream's dynamics can be characterized. If little change is observed over several sample intervals, the test frequency can be extended to reduce the amount of lab work and sampling necessary to obtain reliable test results. It should be noted that this approach can be time consuming and expensive. When budgeting a sampling project, it is important that the cost of this procedure be taken into account so that sufficient funding is available to do the job correctly the first time.

Location of a suitable sampling point (or points) within the waste stream is a critically important consideration in trying to obtain representative samples regardless of whether gases or liquids are being analyzed.

Wastewater should be sampled at or as near the point of discharge as practically possible in order to reliably measure the true effluent conditions at that point. Often, this is a Department of Environmental Conservation (DEC) or EPA requirement.

Wastewater characteristics of primary interest include temperature, pH, chlorine content, turbidity, conductivity (suspended solids), heavy metals, concentrations of other toxic substances, and volumetric flow rate.

pH Level

pH analysis is defined as the measurement of hydrogen-ion concentration in a solution. It is widely used to represent the degree of a solution's acidity or alkalinity.

Acidity or alkalinity of a solution is determined by the ratio of positive hydrogen (H+) ions to negative hydroxyl (H−) ions present. When equal quantities of each are present, the solution is considered to be "neutral." pH level is represented by a range scale of 0 to 14, with a value of 7.0 being neutral. Values above 7 are considered alkaline and values below 7 are acidic.

Table 4-1 identifies the relative pH of common materials found throughout the pH scale.

Nominal Value, pH	Typical Material with this pH Value
14	Caustic soda (4% solution)
13	Caustic hydroxide (lime)
12	Caustic soda (0.04% solution)
11	Ammonia (1.7% solution)
10	Milk of magnesia
9	Borax
8	Sodium bicarbonate
7	Pure water
6	Milk
5	Hydrocyanic acid
4	Beer, orange juice
3	Acetic acid
2	Lemon juice
1	Sulfuric acid
0	Hydrochloric acid (3.7%)
−1	Hydrochloric acid (37%)

Table 4-1. Nominal pH Values

pH measurement is accomplished by electronically sensing the electromechanical potential at the surface of a glass electrode membrane inserted into the sample stream. To accurately measure potential, a reference electrode is also included in the probe assembly to complete the electrical circuit, as shown in Fig. 4-1. The reference electrode maintains electrical contact with the sampled water through a liquid junction between the

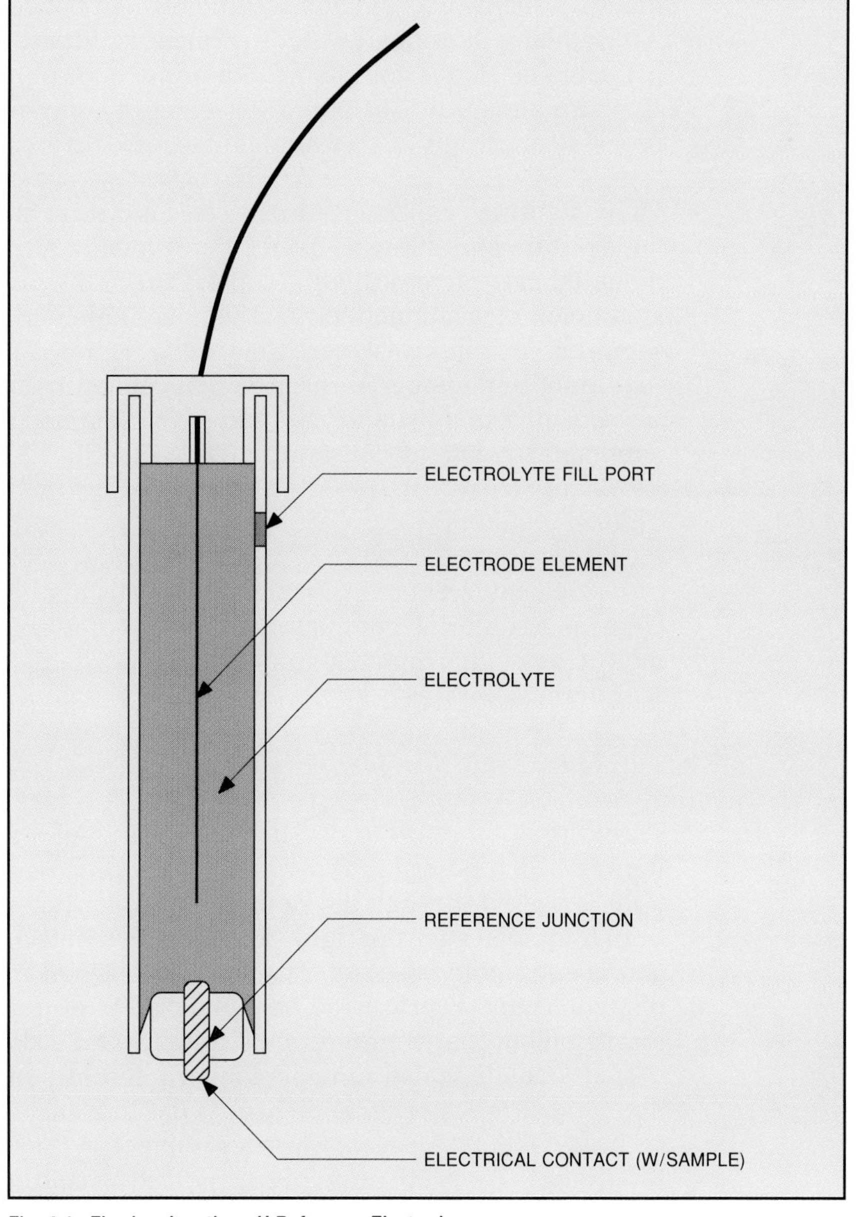

Fig. 4-1. Flowing Junction pH Reference Electrode

electrolyte (usually potassium chloride) fill and the water sample.

By continuously monitoring the upstream pH level in a waste stream, early detection of high or low pH condition is possible, and automatic or manual corrections in water treatment can be made before the wastewater is discharged into the environment.

In most cases, wastewater pH at the point of discharge from a final treatment process should be controlled between 7.0 and 9.0. In many municipal water treatment facilities, an influent pH value of 10.0 is acceptable. An upper control limit of 9.0 provides a margin of safety in the event of a minor upset in process conditions or measurement error.

Many solutions experience increased ionization at higher temperature conditions. The pH measurement electrode's potential also increases with temperature. This can result in significant measurement error if temperature conditions at the sampling location vary more than a few degrees. Therefore, it is important that temperature correction capability be included in the selection of a pH analyzer unless the sampled water temperature remains constant.

Chlorine Content

Chlorine analyzers are available for analysis of either free chlorine or chlorine compounds. Free chlorine consists of molecules of chlorine that are not part of a chemical compound within the solution.

Turbidity

Turbidity is "the ability of a liquid to scatter or attenuate light," which describes the measurement process. A beam of light is passed through the sample liquid. Particles suspended in the liquid scatter the light, which is then detected by optical lenses and photodetectors. The electronic signal from the photodetectors is processed and analyzed to produce a linear measurement that is proportional to the water's clarity. The output value is scaled in nephelometric turbidity units (NTU). The typical range of these devices is 0.1 to 2000 NTU with an accuracy of ±1% of scale for measurement in the 0.1 to 200 NTU range. An NTU value of 0.1 indicates clear liquid (no measurable suspended solids).

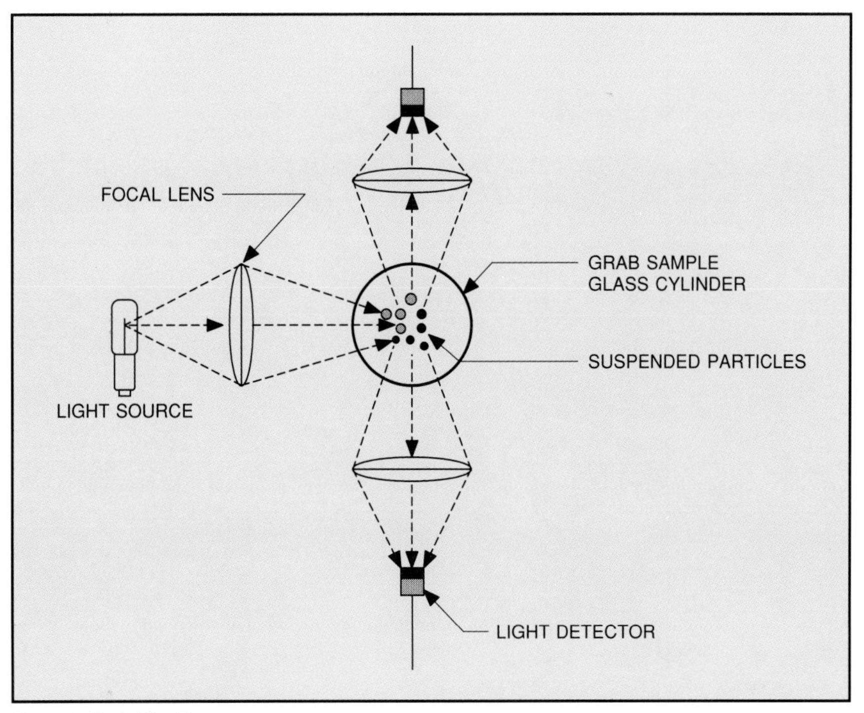

Fig. 4-2. Turbidity Measurement

In continuous measurement applications, a stilling chamber may be required upstream of the sample chamber in order to allow entrapped air within the water sample to escape. It is possible for entrained air bubbles in a turbulent sample stream to disperse the turbidimeter's light beam and cause a significant measurement error.

Turbidimeters are often used for spot sampling, continuous measurement, and laboratory analysis of suspended solids in water samples from rivers, lakes, settling ponds, and lagoons.

Conductivity

The conductivity characteristics of water are used to accurately determine the amount of suspended solids in a sample stream. Measurement is based on the relationship between dissolved contaminants and the resistivity of water (in megohms per cubic centimeter) to electrical current flow. Conductivity, the reciprocal of resistivity, is measured in micromhos or siemens per cubic centimeter, as shown on the graph in Figure 4-3.

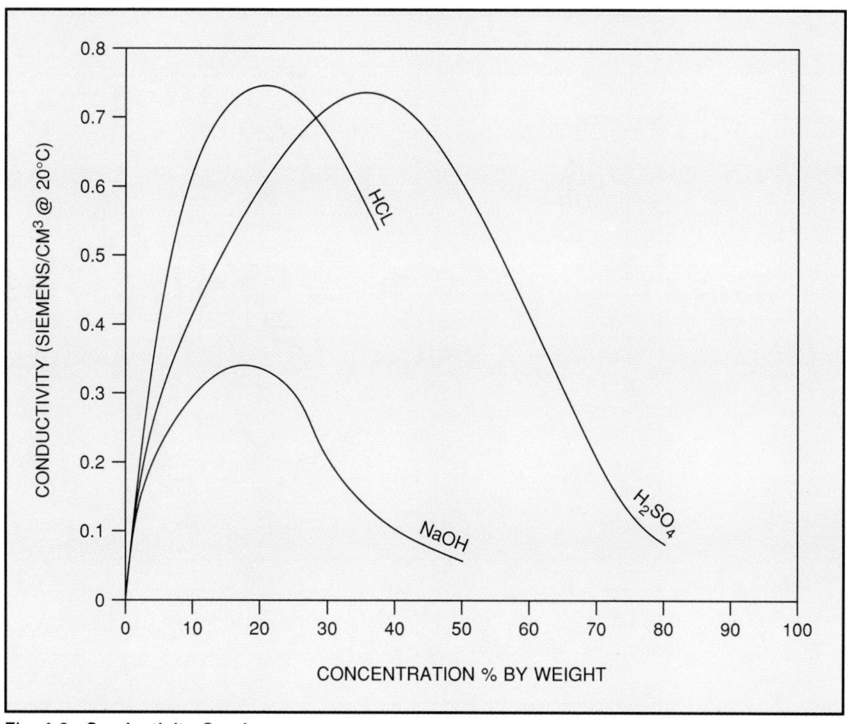

Fig. 4-3. Conductivity Graph

Conductivity is measured by placing a probe containing two electrodes into the sample stream (see Fig. 4-4). Alternating current is applied to the electrodes so that the water completes an electrical circuit. Current flowing through the liquid between the electrodes is measured using a Wheatstone bridge. The output signal from the bridge is then conditioned and amplified to produce a scaled output (4-20 mA DC). Since the sample liquid's temperature can affect measurement accuracy, conductivity analyzers should be equipped with temperature compensation.

Temperature compensation is accomplished by providing a temperature sensor (thermistor or thermocouple) in the sample probe. The characterized signal from the temperature sensor is then used to correct the conductivity analyzer's output accordingly.

4-3. Sampling Systems

Projecting the impact of toxic effluents on the local environment requires accurate measurement of wastewater

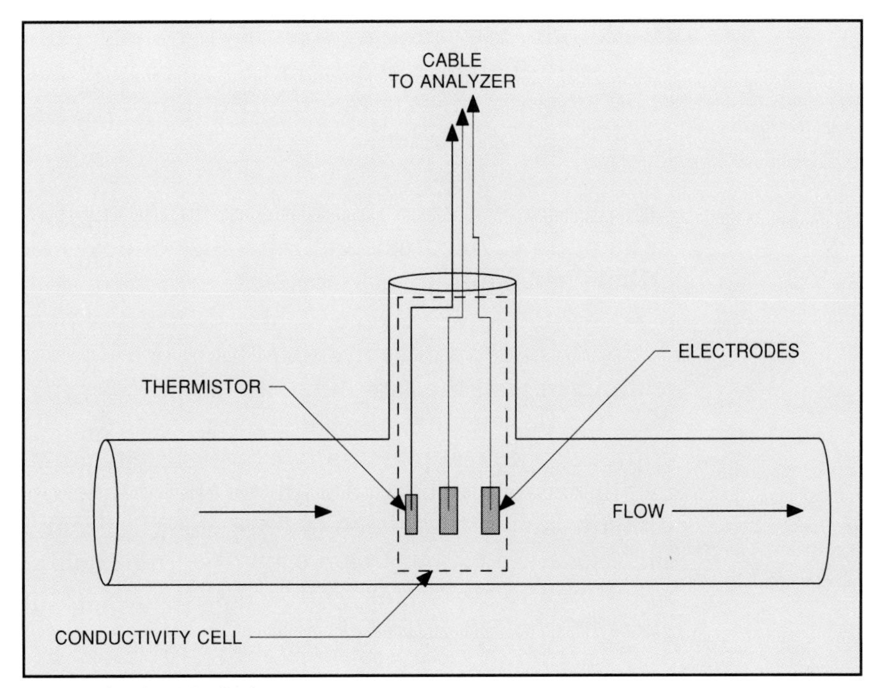

Fig. 4-4. Conductivity Probe

composition and volumetric flow rate. This is accomplished with a continuous sampling program.

Sampling Accuracy

Sampling must be accurate and repeatable in order to assure that the small quantities tested (typically one liter) are representative of what might be a 2-million-gallon-per-day effluent stream. Sampling accuracy for substances, in many cases, must be in the parts-per-billion range. Sampling errors can be very costly to the source facility and the surrounding environment.

Primary factors affecting sampling accuracy are the point within the effluent stream at which sampling takes place and the sampling transport velocity (of the sample). If the point at which samples are drawn does not truly represent an average section of the flow (for example, dead zone, where silt collects) or the transport velocity is slow enough that solids are allowed to settle out, a very significant sampling error will occur. Transport velocity should be at least one foot per second to prevent settling.

In addition, the sampling line inside diameter must be greater than 3/8th-inch to prevent fouling, since most suspended contaminants in wastewater effluent range in sizes of 3/8th inch in diameter or smaller.

Samples are obtained in proportion to the measured rate of effluent flow. Accuracy and reliability of flow measurement are critical to sampling accuracy.

A simple method of sampling effluent is to have a trained technician visit the sampling site periodically to draw samples from the flowing stream. There are several drawbacks to this approach. First, it potentially exposes the technician to any health hazards associated with the effluent. Second, it may be difficult for the technician to draw samples from exactly the same location in the stream each time (repeatability of the sampling process). This approach also takes technician time away from other activities.

Automatic Samplers

Portable and stationary automatic sampling systems are available for temporary and permanent test sites. These systems collect ''grab'' samples from the stream using either a peristaltic pump or a pressure-vacuum system (see Fig. 4-5).

Sample quantities can be selected depending on a predetermined sampling program. Options are available on automatic sampling systems that allow collecting either single or multiple samples in polyethylene or glass container options, ranging from a single five gallon container to 24 mini-containers.

The biggest challenge in automatic sampling is to maintain samples in the same biological and chemical state as they were at the time the sample was obtained. A biological phenomenon known as stasis (a condition in which organisms remain in a relatively fixed state) occurs at or below a temperature of approximately 37°F. For this reason, many stationary (long-term) sampling systems come equipped with a built-in refrigeration system. Portable samplers are often equipped with dry ice, which affords limited control of temperature, since the ice burns off over a period of time and temperature does not remain constant.

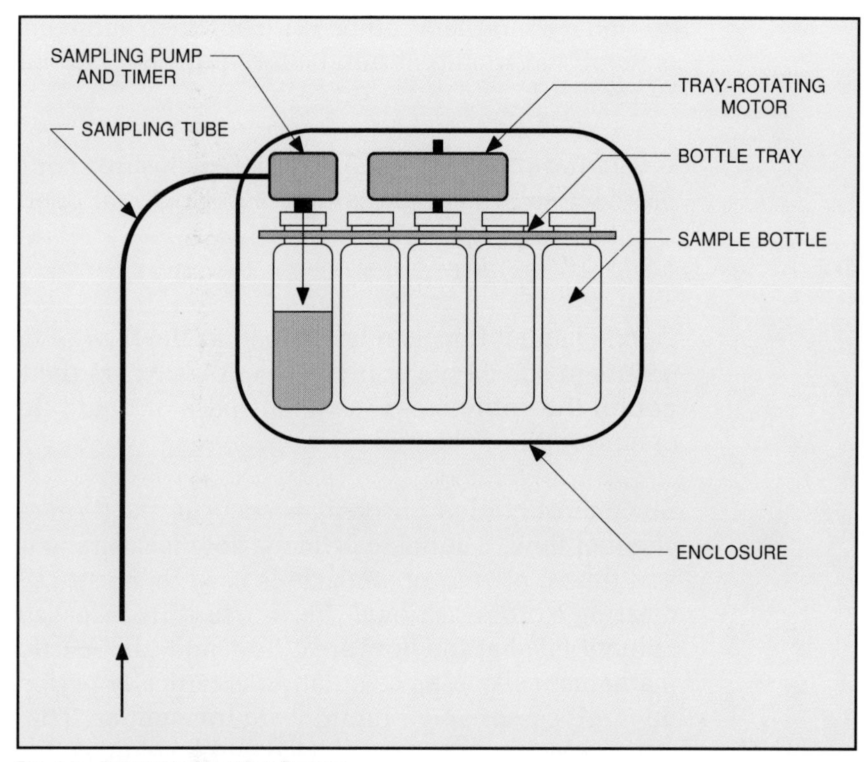

Fig. 4-5. Automatic Sampling System

While mechanical refrigeration systems can be used to maintain a thermostatically controlled environment for the samples, refrigeration compressors tend to draw a significant amount of current and, therefore, require a portable electrical generator or other electrical service in order to operate. Additional battery packs can provide sufficient capacity to sustain operation of refrigeration systems.

Sampling point locations, as emphasized earlier, are a big factor in determining the type of equipment and procedures needed to obtain truly representative samples. A "target list" of contaminants to be measured also impacts these decisions. As an example, volatile organic contaminants (VOCs) cannot be collected using an automatic sampling system because samples are drawn from the effluent stream by producing a partial vacuum in airspace above the sample. Since the compounds are volatile, they turn into vapor within a vacuum and, as a result, are lost in the process.

Stationary samplers can be utilized when sufficient AC power is readily available. If not, battery-powered portable units will be required.

A weather-proof and tamper-proof enclosure should be provided to protect the samplers during their operation.

4-4. Open Channel Flow Measurement

Open channel flow can be defined as the flow of liquid that occurs in some type of pipe or conduit where the liquid does not fill the entire cross-sectional space or liquid flows in an open trough or channel.

Since most effluent discharges occur in the form of open channel flow, a suitable primary flow measurement device such as a flume, a weir, or a nozzle is used to constrict flow slightly, creating a cross-sectional "flow area." By measuring the effluent level at the flow area, flow rate can be calculated mathematically. The calculation is typically performed by a microprocessor and a liquid level transmitter. The combination of primary measurement device and level detection device are referred to as an "open channel flowmeter."

Ideally, flowmeters used for open channel flow measurement should be capable of integrating (totalizing) flow rate, as well as recording and displaying total and instantaneous flow conditions. In addition, they should provide a proportional control signal to the sampler system, thereby allowing automatic adjustment of sampling rate in response to changes in the effluent's volumetric flow rate. When selecting components of the overall system, it is important to select compatible equipment.

Liquid Level Meters

Level measurement devices are available in many forms, including mechanical floats, ultrasonic (non-contact), capacitance, pressure (static head pressure), and bubbler systems (see Fig. 4-6).

The level detection devices gaining in popularity today for open channel level measurement are ultrasonic and capacitance-type level transmitters. Their ease of installation,

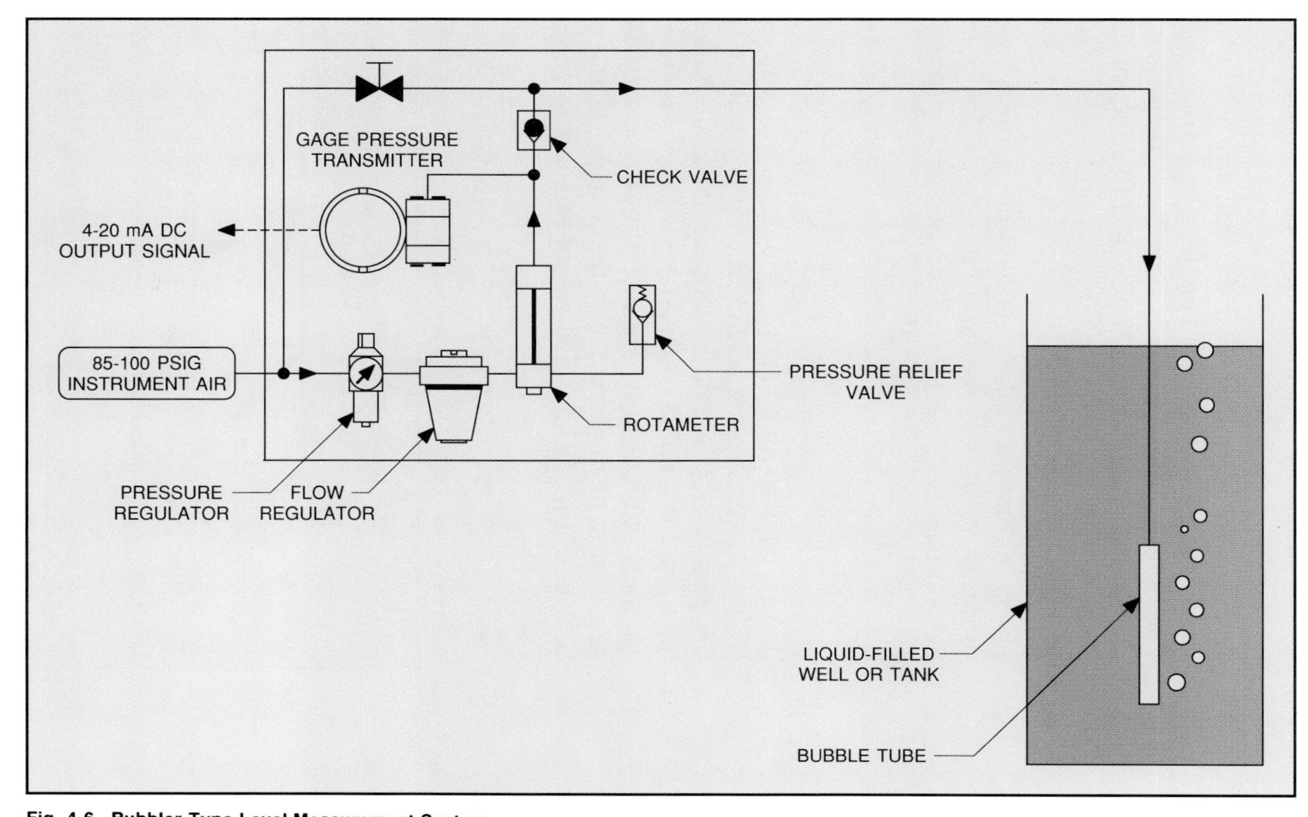

Fig. 4-6. Bubbler-Type Level Measurement System

no mechanical parts, no fouling, low maintenance, and good accuracy make them well-suited for this type of application.

Ultrasonic level measurement devices typically operate at a frequency of approximately 40 kHz. Frequencies higher than this do not transmit as well through air. A 40 kHz sonar pulse in air has a resultant wavelength of approximately 0.33 inch. Therefore, the resolution of Doppler systems is a limit of their transmitted signal wavelength, and the maximum resolution of these devices is $\pm 1/2$ wavelength, or ± 0.16 inch.

The combination of flow element device and level transmitter comprise what is also referred to as a ''velocity-area method'' of measurement.

Weirs

A weir consists of a barrier, or dam, which causes water to collect in a still pond and crest over a precision notch in top of the dam wall. The relationship of the level of the water in the notch and the notch dimensions indicates rate of flow. Weirs are most often used where the effluent has very little material in suspension or floating debris that can settle and collect in the bottom of the still pond ahead of the dam or foul the notch (see Fig. 4-7).

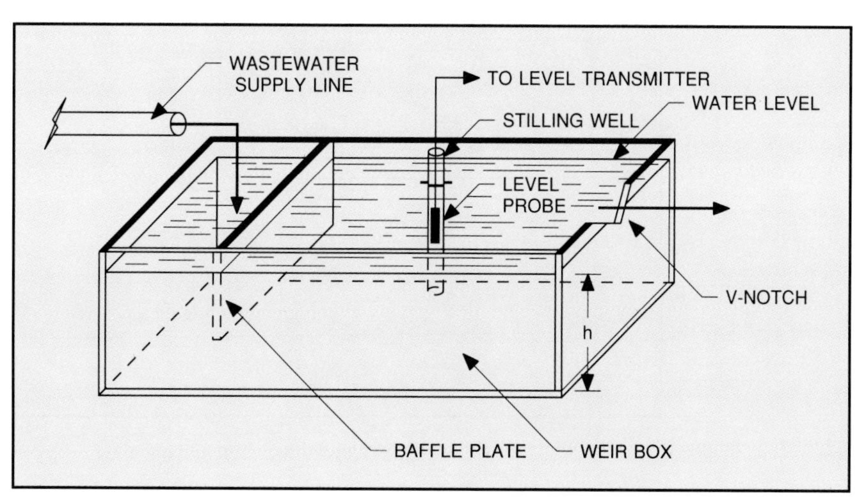

Fig. 4-7. V-Notch Weir

Flumes

A flume is a narrow section of channel shaped to create a critical flow condition. Various flume configurations exist to provide the desired upstream backup of water. The most common forms of flume are the Parshall (see Fig. 4-8) and the Palmer-Bowlus. Parshall flumes are narrower and thus produce a greater backup. This style is most desirable when large quantities of material are held in suspension.

Inaccuracies in flow measurement will occur if a constriction in downstream flow creates a backup of water at the flume or weir's point of level measurement. This can occur due to an inadequate slope or pitch of the channel downstream of the point of measurement, a flow rate that exceeds total capacity of the channel (overflow or flood condition), or downstream blockage. A Parshall flume can accept up to an 80% submergence condition in some cases before the flow rate measurement accuracy is significantly reduced.

Most flumes are sized and selected to operate in the free flow region. However, it is possible to obtain relatively accurate flow rates during submergence when a second level transmitter is installed in the throat (flow area) of the flume. The ratio of downstream level to upstream level is known as the "submergence ratio" and is expressed as a percentage of free flow. With a microprocessor-based system, it is possible to continuously calculate the percentage of submerged flow and

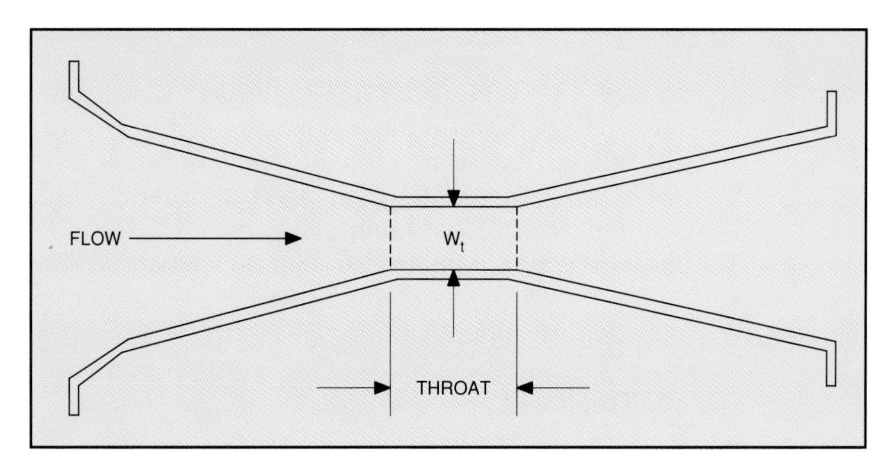

Fig. 4-9. Parshall Flume

use this multiplier to correct transmitted flow value. Obviously, at 100% submergence, no flow will occur.

4-5. Pretreatment of Wastewater

The severity of effluent pollution from any industrial facility depends largely upon characteristics of the processes used to produce an end product, dilution rate of the effluent as it enters the environment, and whether chemical recovery or pretreatment of the effluent occurs prior to its discharge.

Strong public reaction, establishment of stricter government regulations, and enlightened management personnel have all contributed to planning and implementation of in-plant water and chemical reuse (recovery). These actions, in turn, have greatly reduced wastewater quantities and concentrations entering the environment.

In addition to these ''in-plant'' modifications, more facilities are beginning to establish ''out-plant'' waste treatment operations—separate treatment plants to pretreat effluent before it enters the environment (see Fig. 4-9).

A number of control schemes have been developed to regulate the volume and concentration of wastewater effluent discharged from industrial facilities. They typically require the use of pH, conductivity (suspended solids) measurement, and ratio control. In more advanced treatment processes, turbidity, ORP, and residual chlorine are also measured and controlled.

Ratio Control

In a ratio control process, effluent is released into a river or stream (when permissible to do so!) at a flow rate ratioed with the rate of river flow. Effluent enters the river through a submerged, automatic gate valve. In order to do this, the volumetric rate of effluent discharge and stream flow must be measured and compared so that a constant ratio of wastewater to river water is maintained (see Fig. 4-10), in other words, a constant dilution rate.

pH Treatment

Tight control of pH (for example, within 6-8) is not an easy task. Reagent (buffer) demand varies exponentially as pH

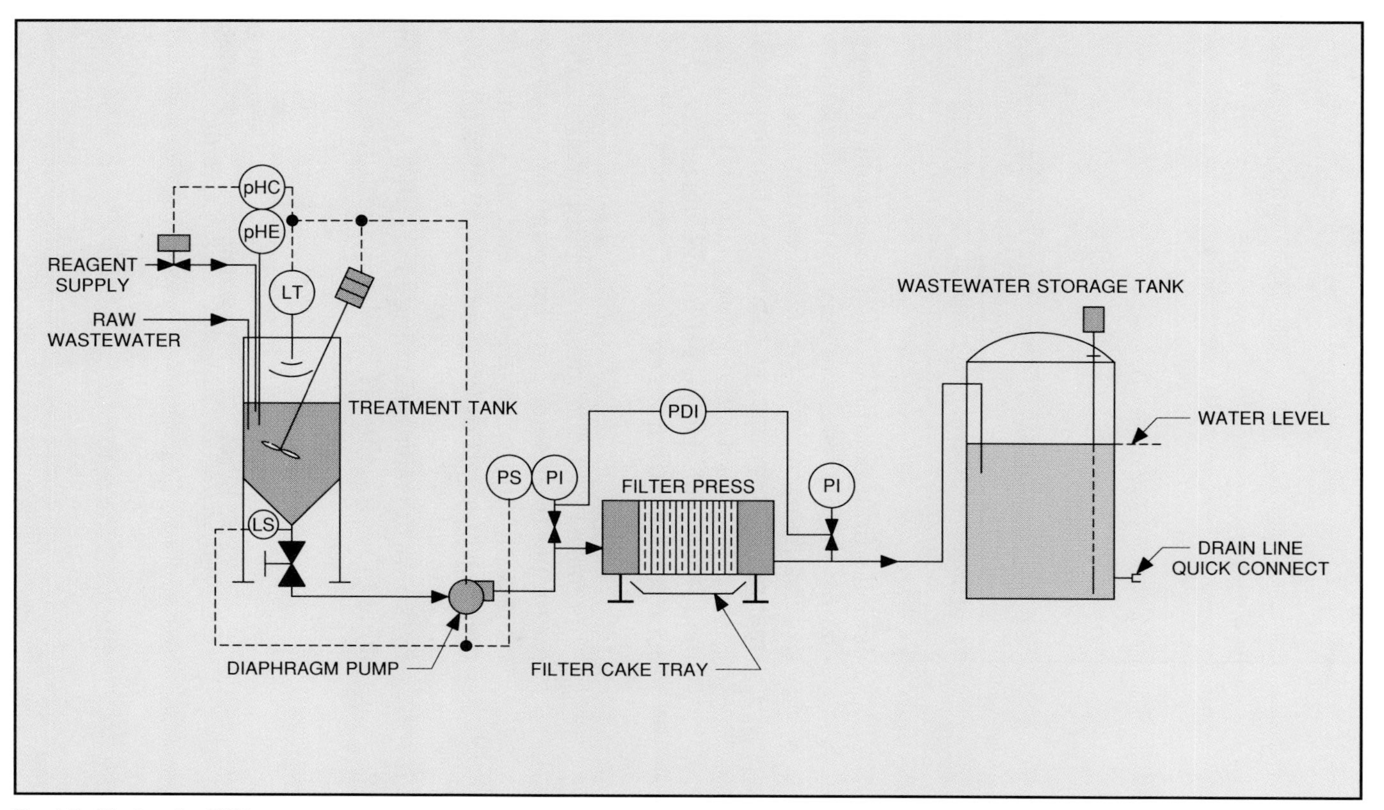

Fig. 4-9. Wastewater P&ID

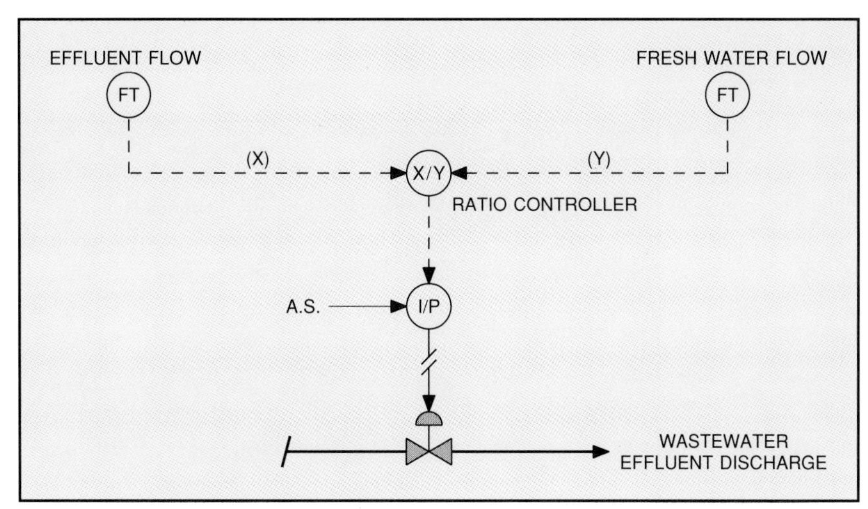

Fig. 4-10. Effluent Discharge Ratio Control

deviates from neutral (7.0). The amount of reagent necessary to change the pH level can vary by a factor of up to 10 million, depending upon how far out of control the pH level is prior to treatment. Therefore, as treatment progresses and pH approaches the desired value (set point), the amount of reagent required to change the pH level changes by a factor of 10 for each level of pH attained (see Table 4-2).

This makes accurate measurement of wastewater's pH critical for proper treatment. Say, for example, wastewater being treated initially has a pH of 9, and the desired pH value is 7. If the measured pH of the control system is off by +1, 10 times the amount of reagent actually needed will be added, resulting in the wastewater becoming too acidic! A base must now be added to bring the pH level back within control range.

pH Level	Deviation	Reagent Demand (10^{-6} moles/liter)
7	0	0
6 or 8	1	1
5 or 9	2	10
4 or 10	3	100
3 or 11	4	1000
2 or 12	5	10,000
1 or 13	6	100,000
0 or 14	7	1,000,000
−1 or 15	8	10,000,000

Table 4-2. Reagent Demand vs. pH Level

pH Measurement

Accurate measurement of pH in a process is difficult. Often, pH sensors are installed improperly to allow accurate measurement. More often, the overall process design (tank and piping configuration) is inadequate. No matter how well the control system performs, it more often than not cannot overcome the poor system dynamics created by inadequate process design.

It is, therefore, very valuable to be able to recognize an inadequate system design before much time and effort are lost in trying to overcome system shortcomings by experimenting with the control system. This situation can often be avoided if the control engineer aids the process engineer during the system's initial design.

Control of pH in a continuous stream is particularly difficult since adequate mixing of the reagent and wastewater may not occur upstream of the measurement location, resulting in poor measurement and control. Therefore, whenever possible, a batch process should be used for wastewater pH treatment. Adequate mixing is then accomplished by injecting reagent within a mixing tank, using either an agitator (mixer) or a recirculation line (see Fig. 4-11). Baffle plates are often added around the interior wall of the tank to increase the mixing action.

pH Process Retention/Dead Time/Gain

Dynamic process characteristics are a function of retention time, dead time, and gain.

Retention time is determined by the volume of a vessel divided by the flow rate through the vessel. The larger the vessel, at a given flow rate, the longer the liquid is retained within the vessel.

Dead time is the time interval between the point at which reagent is added and the time at which a measurable change in pH is detected.

Dynamic gain of the treatment process is based on the formula:

$$G_d = t_o/(6.28t_1)$$

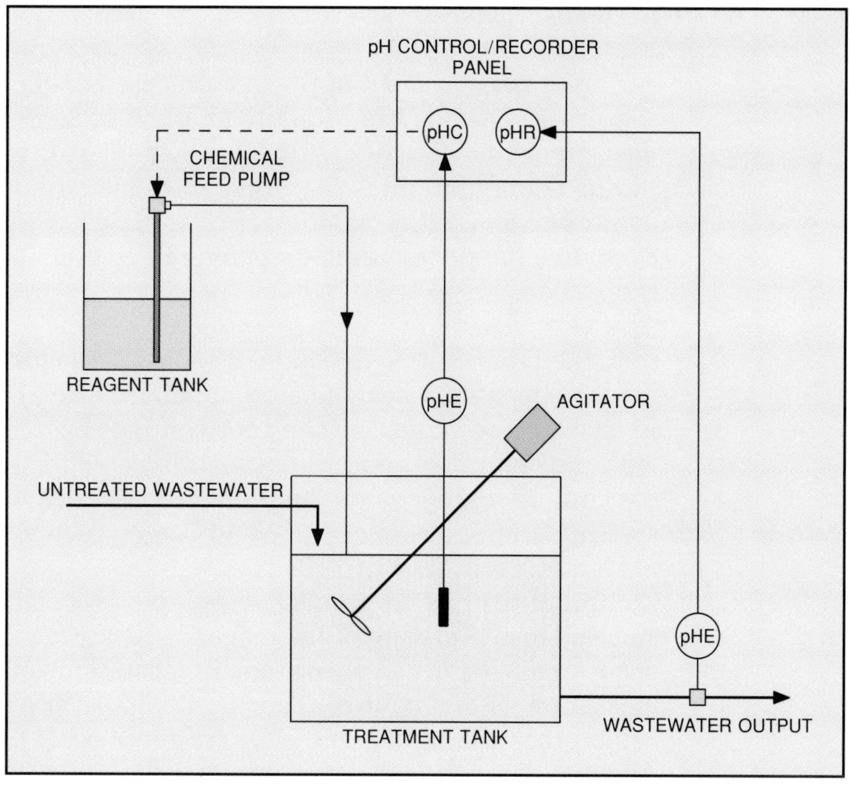

Fig. 4-11. pH Treatment P&ID

where:

G_d = % change in output per % change in input
t_o = disturbance's oscillation time
t_1 = tank retention time (system dead time)

Multiple tanks and mixers piped in series further reduce the overall process gain. This allows the reagent time to react as needed to correct for process upsets (large fluctuations) in incoming wastewater (see Fig. 4-12).

The positioning of agitators and baffle plates within a treatment tank is also quite important. They should be positioned to turn over the contents of the tank rather than create a whirlpool effect. A significant amount of research has gone into the optimum agitator design and its positioning within the tank to produce the most effective mixing of water and reagent. To a large degree, the optimum configuration depends upon the

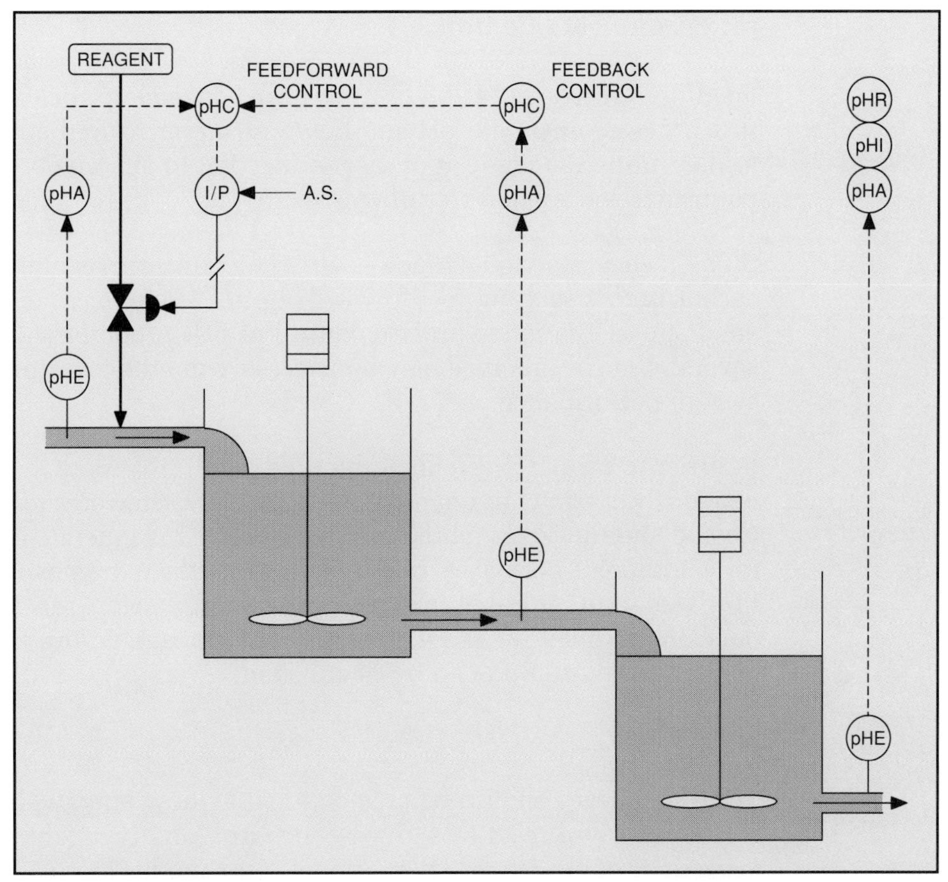

Fig. 4-12. pH Control System

shape and volume of the vessel. To determine the best combination, it is often best to consult the mixer or agitator manufacturer.

To increase retention and dead time within a treatment tank, the tank's inlet and outlet should be on opposite sides of the tank, with the influent entering near the top of the tank and the effluent exiting near the bottom.

A good rule-of-thumb check of the number of vessels recommended for pH treatment is as follows:

No. of Tanks	pH of Influent
1	4 to 10
2	2 to 12
3	<2 to >12

pH Feedforward Control

Feedforward control is often used in pH control applications. In this process, upstream pH and flow rate are measured and used to determine the amount of reagent needed to effectively neutralize the wastewater downstream.

Since measurement accuracy is always a concern, as mentioned earlier in this section, feedforward control alone may not be adequate to guarantee precise control of pH. A combination of several control schemes may be necessary to attain desired system performance.

In order to properly establish the type of pH treatment system needed, variations in characteristics of the wastewater to be treated should be measured and recorded. This must be done for a sufficient period of time to determine the extremes in pH, flow rate, and temperature that can be anticipated. From this data, the number of vessels, the amount of reagent, and the control configuration can be established.

Reagents

Titration curves should be created to show how effectively wastewater will react to a given concentration of reagent. This is easily accomplished when wastewater samples are readily available. In the case of a new waste stream, there may not be an opportunity to pretest the water. In such a situation, if insufficient information is available to make a reasonably accurate calculation, the system should be designed for the worst possible case. Unfortunately, this often results in system overdesign and greater equipment costs. Therefore, if possible, being forced into a worst-case design should be avoided.

Liquid 98% sulfuric acid and 20% sodium hydroxide are common reagents used for pH treatment. These potentially hazardous and highly corrosive solutions must be handled carefully.

4-6. Remedial Systems

Flotation

Wastewater treatment at industrial and municipal facilities often involves a combination of several of the following processes:

A. Chemical recovery

B. Coagulation or flotation (to remove suspended matter)

C. Treatment by biological process, such as activated sludge and trickling filters (to remove oxygen-demanding waste materials)

D. Chemical precipitation (to remove color)

E. Lagooning (for storage, settling)

F. Polishing (for equalization and BOD reduction)

Some constituents of a waste stream are less dense than water and, therefore, float on the water's surface where they can be separated and removed during early stages of pretreatment. This is typically accomplished by pooling wastewater in a large stilling pond where lighter materials are allowed time to float to the surface. A skimming system then collects this material and conveys it to a temporary storage vessel until it can be transported to a disposal site.

Chemical Recovery

This process, much like that used for air emissions applications, involves separation of chemical components from the waste stream prior to its discharge from the plant. Chemical recovery takes many forms, depending largely upon the type of chemicals or compounds being removed.

The most common practice is to force a chemical reaction between the chemical constituent to be removed and an additive, which creates a precipitate that settles out of the stream. The precipitate can then be easily skimmed off or filtered out of the water and disposed of as a sludge material. Filter presses can be used to remove additional filtered water from the sludge. The remaining sludge "cakes" can be handled more easily.

Since the cost to transport waste materials is typically based on weight, it is much more economical to transport relatively dry sludge material than "wet" waste.

Activated Sludge

The activated sludge method is a biological process in which dissolved substances are converted by microorganisms into a form that will settle out of the liquid. This settled material is referred to as "activated sludge."

Settled waste from coagulation is mixed with a portion of the activated sludge and then transferred into an aeration tank. Aeration (introduction of air bubbles) is necessary to provide living bacteria and other microorganisms in the activated sludge with the oxygen they need in order to survive. The aerated material is then transferred to a settling tank, where the activated sludge is separated from the water (see Fig. 4-13). The sludge is further treated and then disposed of. The remaining clear water is then either reused within the facility or discharged to a sewer system or stream if its pH and other attributes are at an acceptable level. If not, further pretreatment is necessary.

Lagooning

Lagoons are large, man-made ponds into which process wastewater is discharged. Lagoons are unlined, allowing most of the wastewater to be absorbed into the surrounding soil.

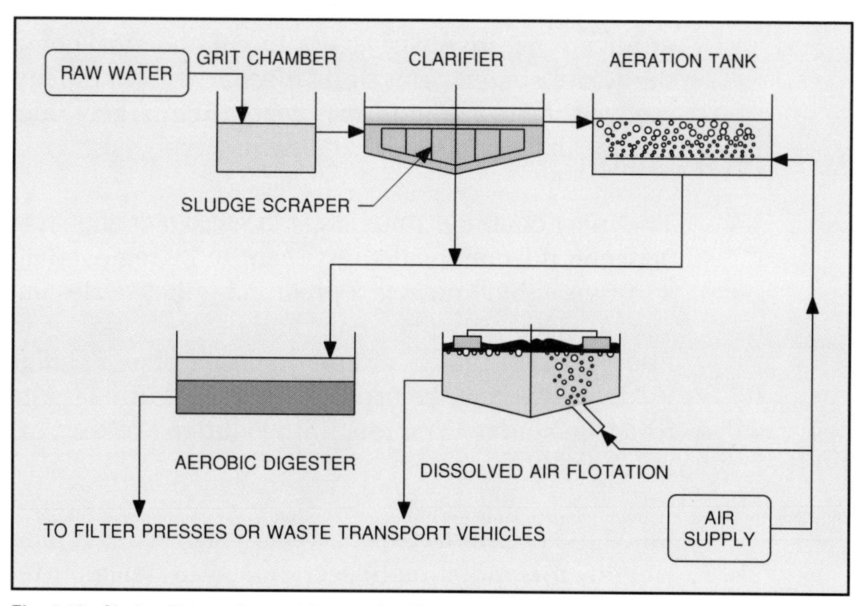

Fig. 4-13. Sludge Thickening and Dewatering Process

Remaining sludge is then scooped out and transported by disposal truck to a sludge disposal facility.

This method of separation is not considered acceptable in most parts of the country if the wastewater contains heavy metals such as lead, mercury, cadmium, nickel, or other toxic materials that can migrate and contaminate surrounding groundwater. Water from nearby monitoring wells is periodically sampled to determine whether this, in fact, is occurring. If so, immediate action must be taken to correct the problem or stiff fines may be incurred by the owner. Acceptance of this type of process should be verified by consulting the regional DEC representative. Pretreatment of wastewater before discharge to the lagoon can remove many, if not all, of the contaminants.

It is likely that future use of lagoons will be limited to wastewater that cannot contaminate groundwater.

References

1. Lavigne, John R., *Instrumentation Applications for the Pulp and Paper Industry*, Miller Freeman Publications (1979).
2. Strauss, Sheldon D., "Water Treatment Control and Instrumentation," *Power*, May, 1988.
3. Hoyle, D. L., "Designing for pH Control," *Chemical Engineering*, November, 1976.
4. Kalis, Gerben, "How Accurate Is Your On-line pH Analyzer?" *INTECH*, June, 1990.
5. Kardos, Paul, "Improving pH Measurement and Control," *Instruments and Control Systems*, April 1981.
6. Cali, Gregory V., "Parameters for Designing a pH Control System," Leeds & Northrup, 1980.
7. Hutchinson, C. William, "Measurement You Can Make or Measurement You Can Use—A Water Treatment Example," *INTECH*, May, 1985.
8. Boyes, Walt, "The How's and Why's of Industrial Effluent Sampling," *Pollution Equipment News*, August, 1990.

Exercises:

4-1. *What is pH? What does it represent?*

4-2. *Why is batch pH treatment of effluent often more effective than continuous flow treatment?*

4-3. *What are the most critical requirements for obtaining accurate, representative samples from a wastewater stream? Where is the ideal sampling location?*

4-4. *What characteristics of wastewater effluent are generally considered to be of greatest interest when monitoring?*

4-5. *What is turbidity, and how is it measured? What purpose is served by a stilling chamber ahead of the turbidimeter?*

4-6. *Why is a refrigeration device or dry ice required for automatic sampling or "grab" samples? When are automatic samplers used?*

4-7. *Define open channel flow. What is a weir? What types of level measurement devices are used for open channel flow measurement? How is flow rate determined by measuring the water level in open channel flow measurement?*

4-8. *Define ratio flow control and the purpose it serves in wastewater emissions.*

4-9. *What determines the amount of reagent that must be added to a wastewater treatment process in order to neutralize it?*

4-10. *Is a wastewater stream with a high pH value considered acidic or alkaline? What type of material is typically added to it in order to lower the pH value to an acceptable level?*

4-11. *What characteristics of pH make it difficult to measure and control? What determines the number of vessels and residence time at each vessel required in a well-designed pH treatment process? Can highly accurate pH instrumentation and controls overcome deficiencies in a pH treatment process?*

4-12. *What is activated sludge and how is it used to treat wastewater? Why is oxygen depletion a problem when using activated sludge?*

4-13. *What are the advantages of chemical recovery systems?*

4-14. *Why are lagoons used to treat effluent? What concerns are associated with the operation of a lagoon?*

Unit 5: Groundwater and Surface Water Monitoring

UNIT 5

Groundwater and Surface Water Monitoring

The fresh water supply is a vital resource. To properly protect it, levels of toxic waste material discharged into it must be closely monitored and controlled. This unit will present the general concepts of groundwater characteristics, analysis, and control methods.

Learning Objectives — When you have completed this unit, you should:

A. Understand the hydrologic cycle and how the fresh water supply migrates.

B. Be familiar with monitoring well construction and the instrumentation used to analyze groundwater conditions.

C. Be familiar with the various monitoring systems designed to detect leaks from underground storage tanks and piping systems.

5-1. Hydrologic Cycle

Groundwater is an element of the hydrologic cycle—Earth's circulatory system that replenishes our fresh water supply. Fresh water is vital to our survival. Water continuously moves between the oceans, air, and land through this cycle (see Fig. 5-1).

Water arrives at a land mass as precipitation in the form of rain, sleet, or snow. A portion of this water returns to the atmosphere through evaporation. Another portion becomes runoff, which enters rivers, lakes, reservoirs, and oceans. The remaining precipitation enters the ground.

Aquifers and Aquitards

Precipitation enters the ground and is either absorbed by plants or continues on deeper into the soil. It passes very slowly through an ''aerated zone'' that consists of porous earth and stone, until it eventually reaches the water table. At or below this point, any porous spots or voids in the soil and/or rock are

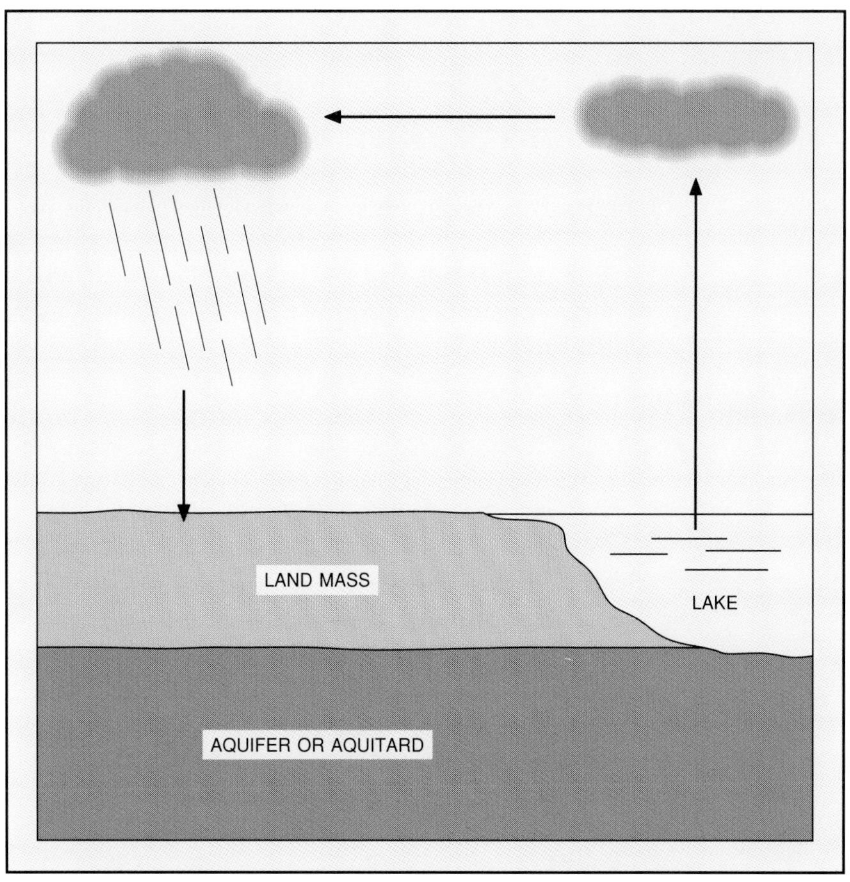

Fig. 5-1. Hydrologic Cycle

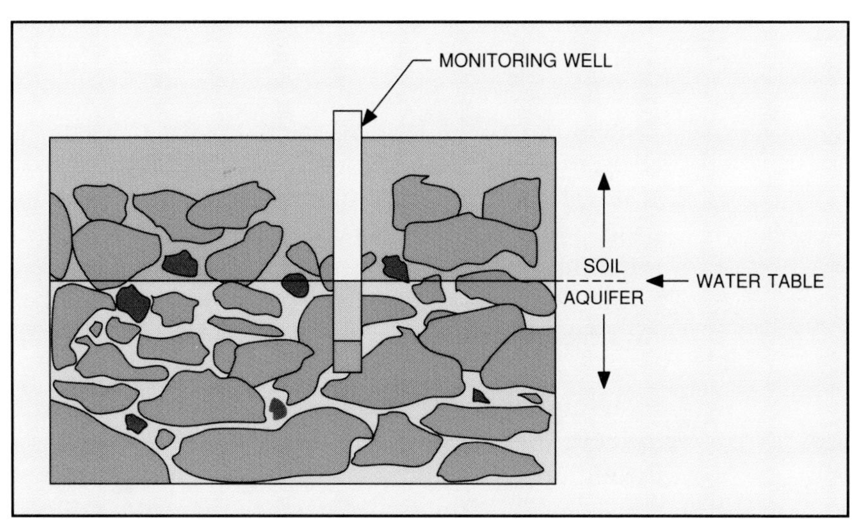

Fig. 5-2. Aquifer

totally saturated. If this saturated zone is capable of providing a significant amount of water to a well or spring, it is commonly referred to as an ''aquifer.'' Aquifers consist of gravel, sand, or porous rock filled with groundwater.

A zone below the water line that does not consist of highly porous material and, therefore, does not contain much water (such as dense rock, shale, or clays) is commonly referred to as an ''aquitard.'' Aquifers and aquitards may be quite small, or they may cover hundreds of square miles below the earth's surface.

Groundwater

Groundwater is constantly acted upon by atmospheric pressure and gravity. The direction in which groundwater propagates is a function of these forces and the surrounding topography. Subsurface groundwater usually flows downhill under the effect of hydraulic gradients. However, this is not always the case. Water will occasionally appear to ''flow uphill'' due to hydraulic pressure created by nearby water displacing it and forcing it upward.

The rate at which water migrates through an aquifer is a function of the soil's or rock's porosity. Groundwater velocity can range anywhere from fractions of an inch to many yards per day. Underground rivers (subsurface channels) exist in relatively rare instances, allowing far more rapid movement of groundwater.

Groundwater, in some instances, will eventually find its way back to the surface in streams, geysers, springs, lakes, or oceans. Depending on its rate of migration, groundwater may be anywhere from a few years to a million years old.

Groundwater levels vary according to the climate, topography, and geological makeup (permeability of the soil or rock layers). In some areas groundwater is found just below grade (surface of the ground), as in the case of wetlands. In other areas, it may be several hundred feet down. Water levels may remain relatively constant year round in some geographical areas, while in others the level will change a number of feet annually.

The closer groundwater is to the surface, the greater the risk of significant contamination by toxic materials released at the surface and the sooner contamination can occur after release of the contaminant.

The continental United States has an estimated 30 to 60 quadrillion gallons of fresh groundwater within 2500 feet of its land surface. This accounts for roughly 96% of the nation's entire fresh water supply and represents more than four times the amount of fresh water contained in the Great Lakes. This source of water is replenished through the hydrologic cycle at an estimated rate of 300 trillion gallons per year.

According to data prepared by the United States Geological Survey (OSGS), about 32 trillion gallons of our nation's groundwater was extracted for use in 1980. About 68% of this water was used for irrigation, 18% for domestic use (drinking, cooking, bathing, lawn care, etc.), 13% for industry, and 1% for livestock.

Sources of Contamination

Spills during the transportation or delivery of toxic liquids can eventually find their way into surrounding soil or water and contribute to the contamination of groundwater.

Materials stored in leaking above- or below-ground tanks can also allow toxic materials to seep into surrounding ground soil and potentially contribute to groundwater contamination.

More specific examples of contaminant sources include leaks from above- and below-ground chemical and petroleum storage tanks, toxic materials collected in rainwater runoff, sewage systems, settling fields and lagoons, landfills (solid waste), and chemical spills.

Some solid materials, such as coal, are simply stored on the surface of the ground. Since this material is not protected from the weather, it is susceptible to a process referred to as "leaching." Rainfall hitting the coal pile gradually washes some of the coal's water-soluble constituents, such as sulfur and ash, into the surrounding soil. Eventually, these chemicals find their way into nearby groundwater.

Leaching can be remedied by constructing a containment system that water-soluble materials can be stored in. This containment system typically consists of a concrete slab on grade with a gutter around the perimeter that catches the runoff. The gutter is pitched, allowing the runoff to collect in a basin at one end. When the basin has partially filled, it is pumped out to a waste disposal vehicle, which then safely transports the wastewater to a treatment facility.

5-2. Monitoring Well Construction

Monitoring wells are constructed to allow technicians to easily reach groundwater for sampling on a periodic or a continuous basis. Manual and automatic sampling systems and level transmitters are often installed at these wells in order to analyze groundwater conditions and makeup.

Often, multiple wells are established at a test site in order to assure that truly representative samples of average groundwater conditions are being obtained. Variations in physical makeup of the aquifer and the surrounding topography will affect the migratory path of groundwater. Therefore, in many cases one well is insufficient for accurate and reliable analysis.

Well construction should also be reliable, yet cost effective. Well shafts must have a large enough inside diameter to allow sample bottles, level monitoring probes, or analyzer probes to be submerged below the groundwater's surface. They must also be sturdy enough to prevent their collapse from surrounding soil pressure and movement. Shafts are often constructed of large sections of 4-inch diameter (or larger) PVC pipe. However, if the potential exists for groundwater contamination by substances that attack PVC (polyvinyl chloride), a substitute material must be considered.

A screen located in the base of the well allows groundwater to easily enter the well until the well's level equals that of the surrounding water table. As mentioned earlier, this depth can be anywhere from a few inches to several hundred feet below the surface.

Obviously, the depth of the water table (top of the aquifer) determines how deep a well must be driven. Topographical information about the area will usually provide some indication of the depth of the water table. Information on any existing nearby wells can also be helpful. Small pilot wells are then drilled to determine the exact depth of groundwater. Wells

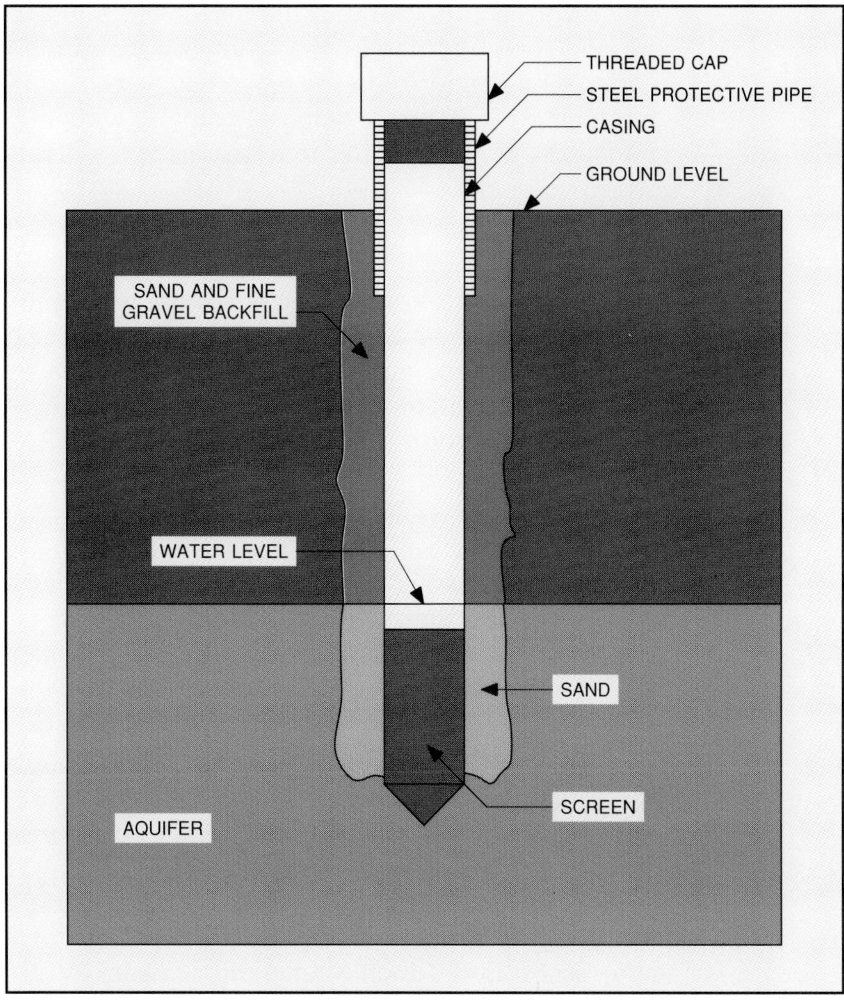

Fig. 5-3. Monitoring Well Construction Detail

must project several feet below the water table in order to allow for seasonal variations in water depth.

System Security

In areas where vandalism can possibly occur, lockable well caps are installed to help discourage vandals from throwing materials into the well that could damage the well or contaminate the groundwater. The greatest potential for vandalism exists in areas where environmental regulations have been blatantly ignored and monitoring efforts may be sabotaged. Any continuous monitoring or sampling systems must also be enclosed and padlocked for protection (see Fig. 5-4).

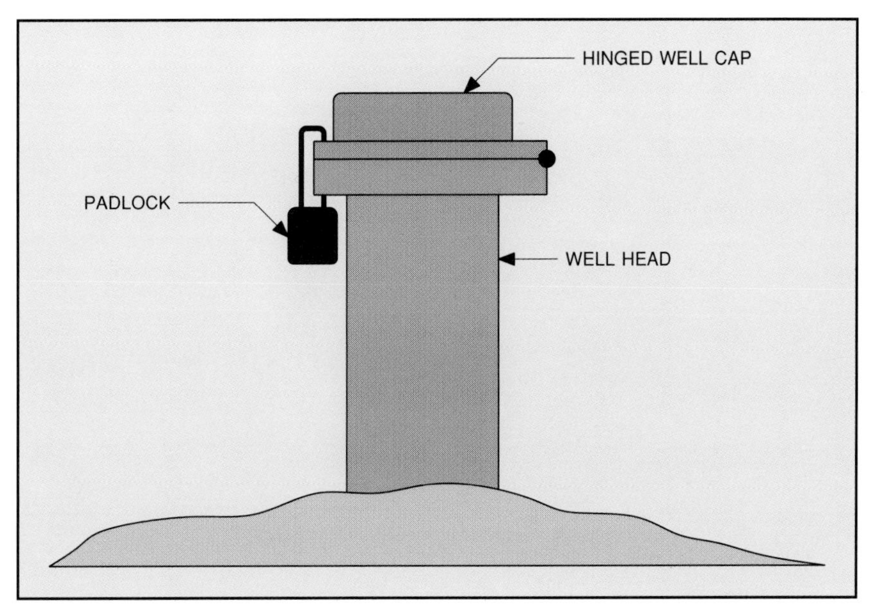

Fig. 5-4. Well Head Protection

5-3. Monitoring Instruments

Groundwater Level Measurement

A variety of instrumentation exists to monitor and record variations in groundwater level. They include such things as manual cable reels, bubbler systems, mechanical floats, and non-contact (ultrasonic) level transmitters (see Figs. 4-6, 5-5 and 5-6). In many cases, groundwater monitoring wells must be established in locations where electrical power or instrument air is unavailable to power continuous measurement and recording equipment. In these remote locations, the engineer or the technician must rely either on portable generators, a solar-powered monitoring system, or battery-powered devices. Ideally, monitoring instrumentation that has low power consumption and can reliably measure and record level changes over many weeks or months without frequent service or maintenance should be used.

Chart Recorders

Traditional chart recorders have several inherent disadvantages when recording level in remote locations for long periods of time:

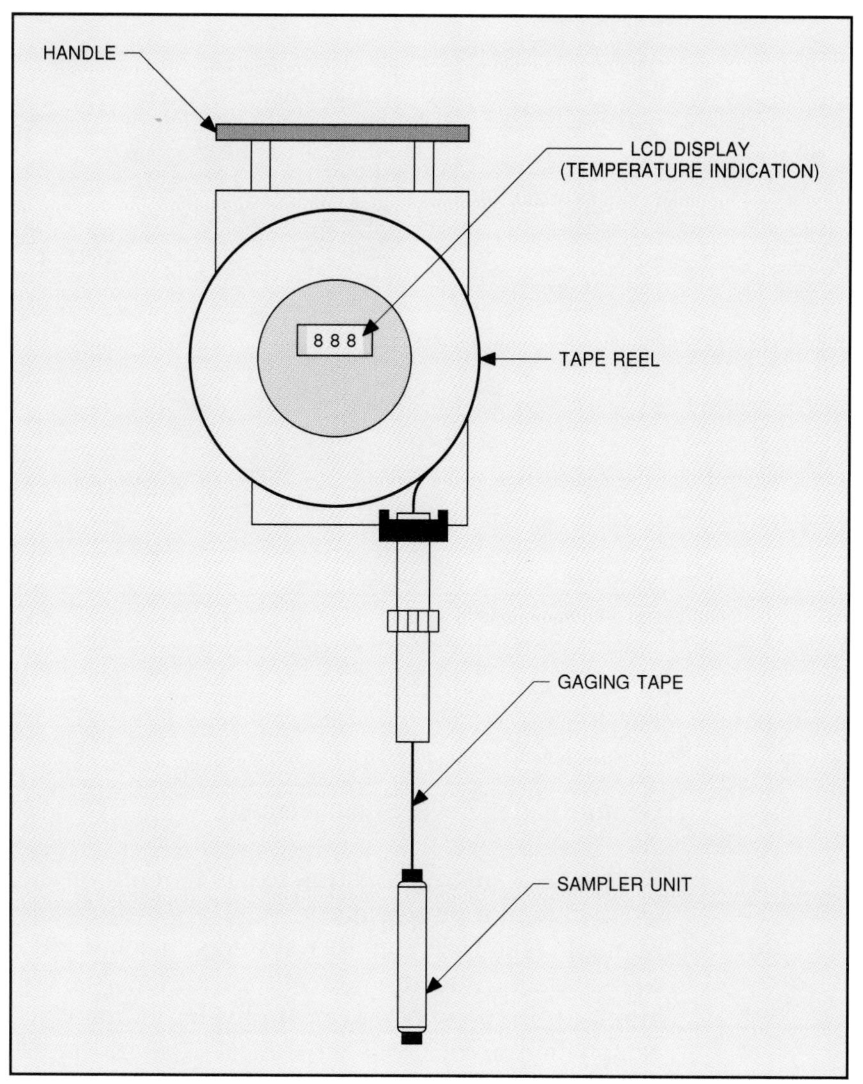

Fig. 5-5. Portable Gaging Tape

A. They require periodic replacement of chart paper and pens.

B. Chart drive motors continuously draw battery current, limiting their operating time when only temporary power is available.

C. Chart recorder paper and pens tend to be very susceptible to humidity and, therefore, must be carefully protected from foul weather.

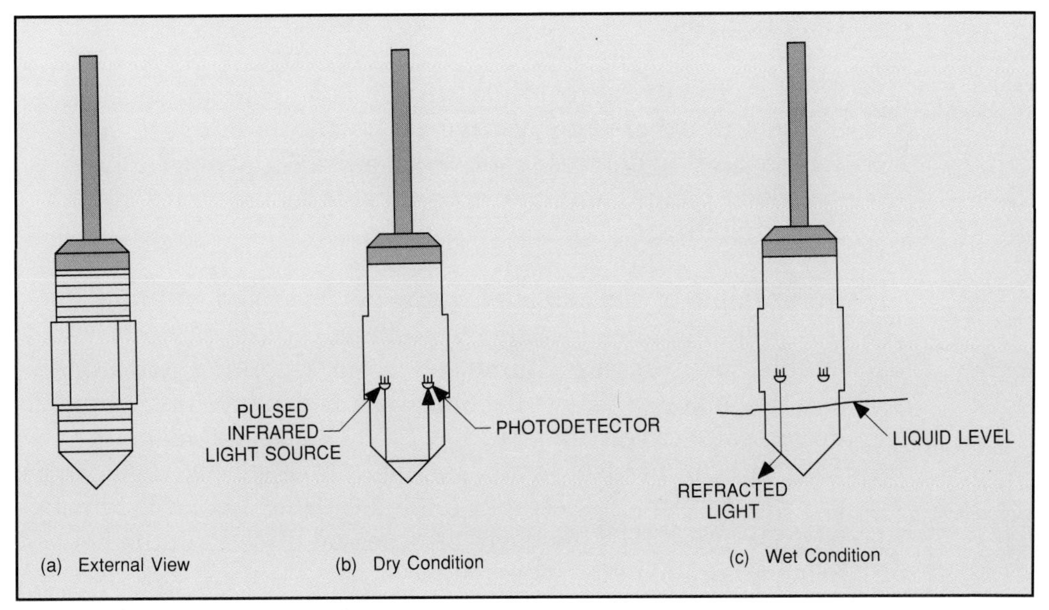

Fig. 5-6. Optical Level Probe Assembly

Circular chart recorders, while very popular for 24-hour or 7-day recording applications, are not well suited for longer recording periods. Slowing the chart speed results in very compressed data and allows ink from the recorder's pens to be absorbed into the paper, making it less precise and legible. This drawback can be partially overcome by available recorders that offer an electronic stylus.

Strip chart recorders are better suited for this type of application. They can be set up to run for weeks at a time by using a long supply of fanfold-type paper. If used in conjunction with an electronic stylus (inkless) versus the traditional fiberous-tip ink pens, servicing is not required for several weeks. By operating the recorder at a low chart speed, recording times can be extended for long periods. Again, slow chart speeds result in some sacrifice of recorded data resolution. This may not be a concern since variations in groundwater level occur very slowly. Surface water conditions, on the other hand, may vary rapidly during a storm and may require greater resolution. Again, much consideration should always be given to the dynamics of the measured variable (in this case water level) when selecting an instrumentation's measurement range and resolution.

Data Loggers

Recent advances in real-time, digital data logging devices have resulted in the creation of compact, relatively low cost, solid-state recording devices that operate at very low power levels over long periods and offer considerable accuracy, resolution, and durability.

An example of such a data logging device is shown in Fig. 5-7. It is preprogrammed for the type of analysis (level, flow, temperature, pressure, humidity), scale, recording start time, and frequency of data collection. Data is displayed using menu-driven software on an IBM PC/XT™,AT™, or PS2™ computer with 512K or more of RAM (see Fig. 5-8). Recorded data can be transferred from the recorder to the PC using a hand-held data transfer unit (DTU), or it can be accessed directly in the field using a portable PC.

This data logger can be directly connected to a desktop PC anywhere convenient for initial programming and then

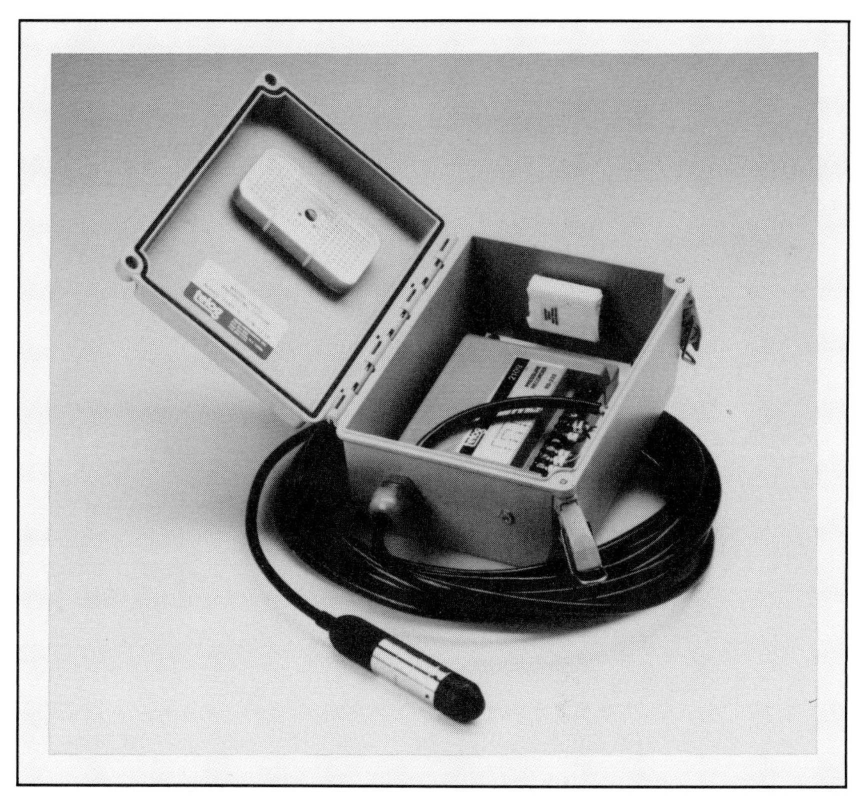

Fig. 5-7. Data Logging Device (Courtesy of Telog Instruments, Inc.)

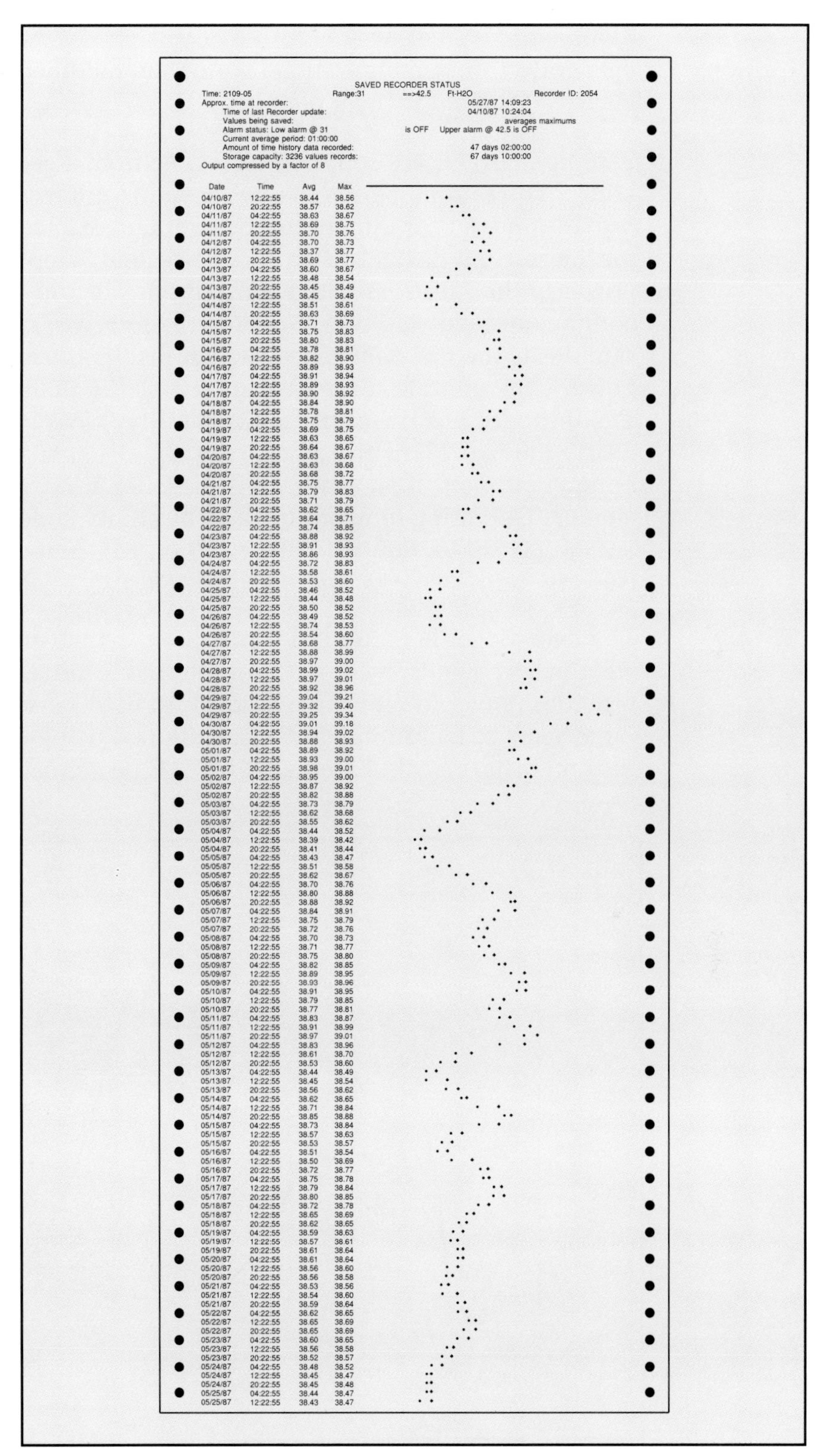

SAVED RECORDER STATUS

Time: 2109-05 Range:31 ==>42.5 Ft-H2O Recorder ID: 2054
Approx. time at recorder: 05/27/87 14:09:23
Time of last Recorder update: 04/10/87 10:24:04
Values being saved: averages maximums
Alarm status: Low alarm @ 31 is OFF Upper alarm @ 42.5 is OFF
Current average period: 01:00:00
Amount of time history data recorded: 47 days 02:00:00
Storage capacity: 3236 values records: 67 days 10:00:00
Output compressed by a factor of 8

Date	Time	Avg	Max
04/10/87	12:22:55	38.44	38.56
04/10/87	20:22:55	38.57	38.62
04/11/87	04:22:55	38.63	38.67
04/11/87	12:22:55	38.69	38.75
04/11/87	20:22:55	38.70	38.76
04/12/87	04:22:55	38.70	38.73
04/12/87	12:22:55	38.37	38.77
04/12/87	20:22:55	38.69	38.77
04/13/87	04:22:55	38.60	38.67
04/13/87	12:22:55	38.48	38.54
04/13/87	20:22:55	38.45	38.49
04/14/87	04:22:55	38.45	38.48
04/14/87	12:22:55	38.51	38.61
04/14/87	20:22:55	38.63	38.69
04/15/87	04:22:55	38.71	38.73
04/15/87	12:22:55	38.75	38.83
04/15/87	20:22:55	38.80	38.83
04/16/87	04:22:55	38.78	38.81
04/16/87	12:22:55	38.82	38.90
04/16/87	20:22:55	38.89	38.93
04/17/87	04:22:55	38.91	38.94
04/17/87	12:22:55	38.89	38.93
04/17/87	20:22:55	38.90	38.92
04/18/87	04:22:55	38.84	38.90
04/18/87	12:22:55	38.78	38.81
04/18/87	20:22:55	38.75	38.79
04/19/87	04:22:55	38.69	38.75
04/19/87	12:22:55	38.63	38.65
04/19/87	20:22:55	38.64	38.67
04/20/87	04:22:55	38.63	38.67
04/20/87	12:22:55	38.63	38.68
04/20/87	20:22:55	38.68	38.72
04/21/87	04:22:55	38.75	38.77
04/21/87	12:22:55	38.79	38.83
04/21/87	20:22:55	38.71	38.79
04/22/87	04:22:55	38.62	38.65
04/22/87	12:22:55	38.64	38.71
04/22/87	20:22:55	38.74	38.85
04/23/87	04:22:55	38.88	38.92
04/23/87	12:22:55	38.91	38.93
04/23/87	20:22:55	38.86	38.93
04/24/87	04:22:55	38.72	38.83
04/24/87	12:22:55	38.58	38.61
04/24/87	20:22:55	38.53	38.60
04/25/87	04:22:55	38.46	38.52
04/25/87	12:22:55	38.44	38.48
04/25/87	20:22:55	38.50	38.52
04/26/87	04:22:55	38.49	38.52
04/26/87	12:22:55	38.74	38.53
04/26/87	20:22:55	38.54	38.60
04/27/87	04:22:55	38.68	38.77
04/27/87	12:22:55	38.88	38.99
04/27/87	20:22:55	38.97	39.00
04/28/87	04:22:55	38.99	39.02
04/28/87	12:22:55	38.97	39.01
04/28/87	20:22:55	38.92	38.96
04/29/87	04:22:55	39.04	39.21
04/29/87	12:22:55	39.32	39.40
04/29/87	20:22:55	39.25	39.34
04/30/87	04:22:55	39.01	39.18
04/30/87	12:22:55	38.94	39.02
04/30/87	20:22:55	38.88	38.93
05/01/87	04:22:55	38.89	38.92
05/01/87	12:22:55	38.93	39.00
05/01/87	20:22:55	38.98	39.01
05/02/87	04:22:55	38.95	39.00
05/02/87	12:22:55	38.87	38.92
05/02/87	20:22:55	38.82	38.88
05/03/87	04:22:55	38.73	38.79
05/03/87	12:22:55	38.62	38.68
05/03/87	20:22:55	38.55	38.62
05/04/87	04:22:55	38.44	38.52
05/04/87	12:22:55	38.39	38.42
05/04/87	20:22:55	38.41	38.44
05/05/87	04:22:55	38.43	38.47
05/05/87	12:22:55	38.51	38.58
05/05/87	20:22:55	38.62	38.67
05/06/87	04:22:55	38.70	38.76
05/06/87	12:22:55	38.80	38.88
05/06/87	20:22:55	38.88	38.92
05/07/87	04:22:55	38.84	38.91
05/07/87	12:22:55	38.75	38.79
05/07/87	20:22:55	38.72	38.76
05/08/87	04:22:55	38.70	38.73
05/08/87	12:22:55	38.71	38.77
05/08/87	20:22:55	38.75	38.80
05/09/87	04:22:55	38.82	38.85
05/09/87	12:22:55	38.89	38.95
05/09/87	20:22:55	38.93	38.96
05/10/87	04:22:55	38.91	38.95
05/10/87	12:22:55	38.79	38.85
05/10/87	20:22:55	38.77	38.81
05/11/87	04:22:55	38.83	38.87
05/11/87	12:22:55	38.91	38.99
05/11/87	20:22:55	38.97	39.01
05/12/87	04:22:55	38.83	38.96
05/12/87	12:22:55	38.61	38.70
05/12/87	20:22:55	38.53	38.60
05/13/87	04:22:55	38.44	38.49
05/13/87	12:22:55	38.45	38.54
05/13/87	20:22:55	38.56	38.62
05/14/87	04:22:55	38.62	38.65
05/14/87	12:22:55	38.71	38.84
05/14/87	20:22:55	38.85	38.88
05/15/87	04:22:55	38.73	38.84
05/15/87	12:22:55	38.57	38.63
05/15/87	20:22:55	38.53	38.57
05/16/87	04:22:55	38.51	38.54
05/16/87	12:22:55	38.50	38.69
05/16/87	20:22:55	38.72	38.77
05/17/87	04:22:55	38.75	38.78
05/17/87	12:22:55	38.79	38.84
05/17/87	20:22:55	38.80	38.85
05/18/87	04:22:55	38.72	38.78
05/18/87	12:22:55	38.65	38.69
05/18/87	20:22:55	38.62	38.65
05/19/87	04:22:55	38.59	38.63
05/19/87	12:22:55	38.57	38.61
05/19/87	20:22:55	38.61	38.64
05/20/87	04:22:55	38.61	38.64
05/20/87	12:22:55	38.56	38.60
05/20/87	20:22:55	38.56	38.58
05/21/87	04:22:55	38.53	38.56
05/21/87	12:22:55	38.54	38.60
05/21/87	20:22:55	38.59	38.64
05/22/87	04:22:55	38.62	38.65
05/22/87	12:22:55	38.65	38.69
05/22/87	20:22:55	38.65	38.69
05/23/87	04:22:55	38.60	38.65
05/23/87	12:22:55	38.56	38.58
05/23/87	20:22:55	38.52	38.57
05/24/87	04:22:55	38.48	38.52
05/24/87	12:22:55	38.45	38.47
05/24/87	20:22:55	38.45	38.48
05/25/87	04:22:55	38.44	38.47
05/25/87	12:22:55	38.43	38.47

Fig. 5-8. Hard Copy Printout (Courtesy of Telog Instruments, Inc.)

mounted in a small weatherproof enclosure near the well for data collection. Configuration or program modifications can be made in the field using a portable PC.

Sample intervals are software selectable from 1 second up to 8 hours. The sample period is one second. An internal battery provides power for the microprocessor and excitation voltage for the sensing device. After the one second sampling time, power to the sensor and logic is shut off. The only system component requiring power during the interval between samples is the real-time clock, which results in extremely low power consumption. Storage memory is 8K (6500 data values). The unit also has linear scaling capability. Measurement resolution is 0.1% with an accuracy of ±0.2% of full scale. Temperature effect on accuracy is 0.1% for a 5 degree Celsius change. Operating temperature and humidity conditions are −25°C (13°F) to +60°C (140°F) at 95% RH, non-condensing (see Fig. 5-9).

The data logger is housed in a 4.5 × 6 × 1-inch enclosure and weighs less than two pounds. An internal lithium battery provides power to the data recorder and field sensor. Up to three years of measurement can be attained using a secondary lithium battery pack. External 120 V AC power can also be used.

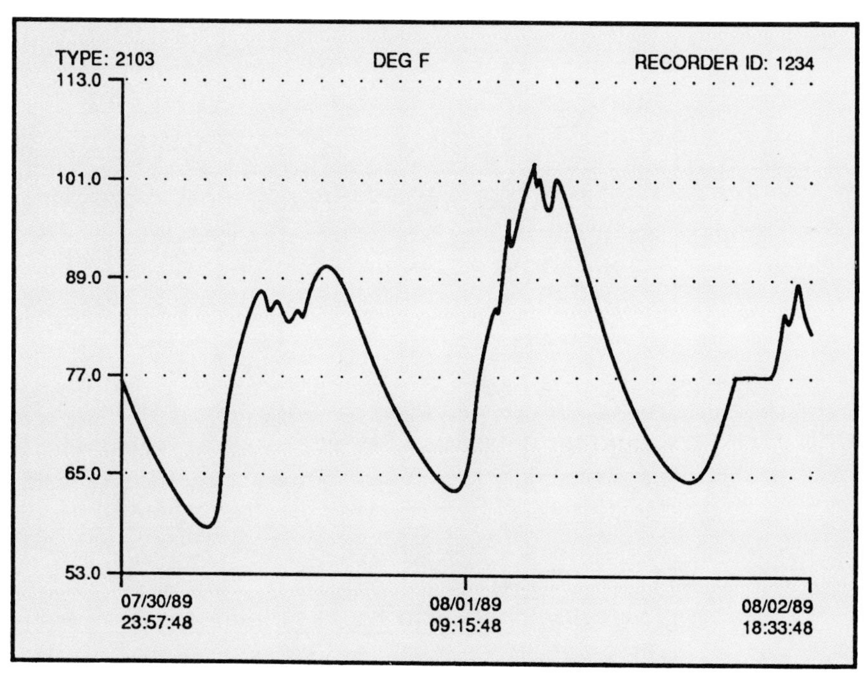

Fig. 5-9. Data Logger Graph (Courtesy of Telog Instruments, Inc.)

Three-input versions of this device are available to accept inputs from float/counterweight/potentiometer assemblies, a nitrogen bubbler system using pressure transducers, or submersible strain gage pressure transducers for level measurement. These measurement systems are typically housed in a NEMA 3R (rainproof) enclosure along with secondary lithium battery packs for extended logging time.

An advanced version of the standard recorder is capable of monitoring and recording up to four input variables and contains a built-in modem for remote monitoring and recording when phone lines are accessible.

5-4. Sampling Systems

The simplest and oldest form of groundwater sampling is by manually obtaining a (''grab'') sample from surface water or a monitoring well (weighted sample bottle is lowered into the monitoring well and fills with water). However, this method may require frequent visits to the well and can expose the sampling technician to toxic materials. If this situation is unavoidable, the technician should be equipped with personal protection equipment (safety glasses, respirator, protective clothing, etc.).

Construction of monitoring wells is not always required to sample subsurface water. Manual sampling can also be performed, when the water table is near the surface, by using a hand-operated boring instrument. This device has a point at one end and is constructed of tubular steel. The device is first driven into the ground, and the outer wall of the tube is retracted slightly to provide a small opening near the nose of the device. A vacuum connection is made at the top of the device, and the groundwater is drawn through the device and collected in sampling bottles for later analysis.

Automatic sampling systems perform the same task as grab sampling with less labor and in most cases less potential exposure. These systems are particularly useful in remote locations where frequent visits to the sampling site would be very time-consuming and costly.

The automated sampling system uses a small electric pump to lift a measured quantity of water from the well through a flexible hose and fill a sample bottle. Once the sample is obtained, the bottle tray is rotated until the next sample bottle is in position to be filled (see Fig. 4-5).

A programmable timer determines when the next sample is to be drawn from the well. The smaller the sample required by the laboratory for analysis, the more bottles that can be fit into the sampling rack. If more samples are required than will fit into the sampler, a second sampler can be set up at the same location and programmed to begin collecting samples after the first unit has obtained a complete set. Samples are refrigerated until collected for lab analysis.

As mentioned in Unit 4, the biggest challenge of any automatic sampling system is maintaining samples in the same biological and chemical state as they were at the time the sample was obtained. Stasis occurs at or below a temperature of approximately 37°F (3°C). Stationary sampling systems are equipped with built-in refrigeration. Portable samplers are often packed with dry ice, affording limited, temporary control of temperature, since the ice burns off over a period of time. As a result, sample temperatures do not remain constant. For limited use, however, this method can maintain samples below 37°F.

If electrical power is unavailable at the sampling site, sampling systems are powered by battery packs and ice is used to refrigerate the samples.

On-Line Continuous Water Monitoring

Portable analyzers capable of providing continuous monitoring and analysis of water conditions are also available when quick analysis is required. This is particularly important where water conditions change rapidly and could create immediate health or environmental risks.

The portable/continuous system shown in Fig. 5-10 operates on the principle of "purge and trap chromatography." The unit is equipped with an adsorption/desorption trap and gas chromatograph operated through a lap-top personal computer. An oven within the unit houses a sample column and detector.

Detectors for the unit are selected based upon the type of substances to be analyzed. Argon ionization is used for general-purpose detection of organic compounds. Electron capture is

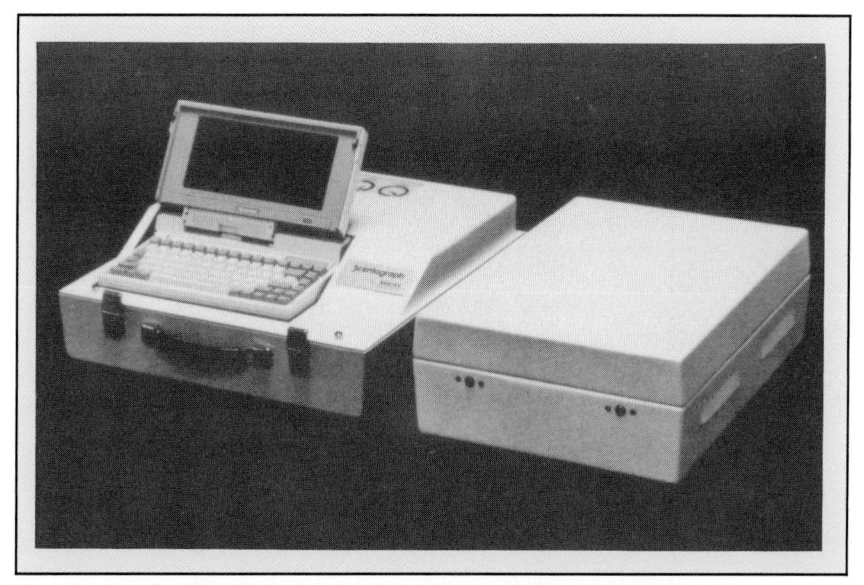

Fig. 5-10. Portable Chromatograph (Courtesy of Sentex Sensing Technology, Inc.)

used for analysis of halogenated compounds. Photoionization may also be used for analyzing organics.

Vapors from the water sample are carried into the trap and desorbed into the chromatographic column for separation and detection. Then a chromatogram is displayed on the computer while retention times and peak areas are recorded in memory and printed out (see Fig. 5-11). Once the sample has been analyzed, it is flushed from the system with distilled water to avoid contamination.

The system can also be programmed to activate an alarm if the preselected concentration levels exceed their programmed limit.

Frequency of analysis is programmed by the operator. Up to 20,000 chromatographic traces (see Fig. 5-11) can be stored on the 20 megabyte hard drive.

5-5. Underground Storage Tank Monitoring

Underground storage tanks (USTs) have been used for many years to store chemicals and petroleum products. Chemicals regulated by the EPA's UST standards are identified within the Comprehensive Environmental Response, Compensation, and

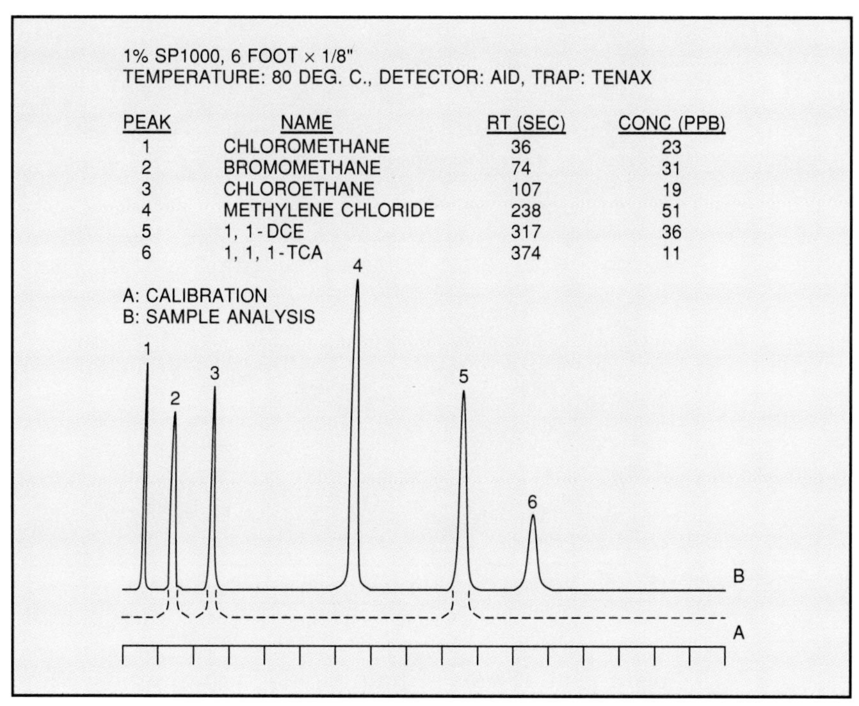

Fig. 5-11. Chromatogram of Wastewater Stream (Courtesy of Sentex Sensing Technology, Inc.)

Liability Act (CERCLA) of 1980 (also known as the Superfund Act).

Underground storage systems are often used because of limited storage space above ground, fire protection (in the case of highly flammable materials), or for esthetics, since above-ground tanks rarely add to a facility's appearance.

Since USTs are buried, there is no way, short of digging them up, to visually inspect for leaks. For years, unless there was a catastrophic failure of the tank or custody of the material (liquid delivered versus liquid removed) was disputed, there was often no way to know whether a tank was leaking (or how much!) until the liquid or its odor showed up in the drinking water or surface water somewhere nearby and was traced.

The EPA has estimated that as many as 30% of all underground storage tanks in the U.S. presently leak. The expense associated with the cleanup of underground tank spills averages about $100,000 per incident. Federal regulations also mandate that owners of the tanks be insured for $1 million per incident, including cleanup and compensation for injury or property damage. Therefore, regardless of the fact that monitoring of underground storage systems is a government regulation, the cost of installing a leak detection system is less than the expense of most cleanup efforts.

Another alternative is to replace below-ground tanks with new above-ground tanks. Doing so requires installation of a secondary containment system to prevent leaks from the above-ground tank from entering the ground. Containment systems must be designed in such a manner that the entire contents of the tank could spill and not overrun the containment barrier.

Typical above-ground containment systems often consist of a sealed concrete basin or earthen barrier with a clay or nonpermeable liner. The regional DEC representative should be consulted when such a system is being designed. Local requirements vary.

While above-ground tanks can often be inspected more easily than below-ground systems, a properly designed and installed underground system can also safely and reliably store materials.

New underground storage tanks must be constructed of fiberglass, cathodically protected double-wall steel, or fiberglass-lined steel.

Provisions for new tank installations must also include overfill alarms and spill containment at the tank's loading/unloading station.

Several methods of leak detection monitoring have been approved by the EPA. These include automatic tank gaging, soil monitoring, groundwater monitoring, a monthly inventory accompanied by an annual tank-tightness test, and monitoring of the interstitial space between the inner and outer tank walls.

Underground Piping Systems

As of December, 1990, EPA regulations require all underground pressurized piping systems be monitored and equipped with an automatic shutoff device in the event of a leak (see Table 5-1). Instrumentation used for leak detection must be capable of detecting a leakage rate of 3 gallons per hour or less at 10 psig.

UST Monitoring Method (Monthly Monitoring)	Existing Tanks	New Tanks
Automatic Tank Gaging Systems	Yes	Yes
Vapor/Soil Monitoring Systems	Yes	Yes
Groundwater Monitoring Systems	Yes	Yes
Monitoring of Double-wall Tank Interstitial Space	Yes	Yes
Monthly Tank Inventory/ Annual Tank-Tightness Tests	(thru 1998)	(10 yr. limit)

Table 5-1. EPA-Approved Underground Storage Tank Monitoring Methods

All new installations must be equipped with this system at the time of construction. In addition, monthly leak detection tests or an annual tightness test must be conducted.

Leak detection of negative-pressure (suction) piping must be monitored monthly or tightness tested every three years.

The monitoring of piping systems may be accomplished by interstitial sensors, vapor detection, groundwater monitoring wells, or annual precision testing.

Line pressure detection can be used on single-wall, pressurized piping systems. This method of detection operates on the premise that a loss in line pressure indicates a leak occurring somewhere in the piping system. A pressure switch or series of switches installed in threaded "T" connections along the containment pipe system act as an alarm contact input that activates a remote alarm notification panel of a system failure upon loss of pipeline pressure. This method can be susceptible to fluctuations in line pressure caused by conditions such as a loss of pump pressure or flow restrictions. Momentary fluctuations can be overcome by designing a sufficient time delay on alarm activation to avoid nuisance alarms.

It should be mentioned at this point that nuisance alarms created by any type of monitoring system quickly reduce the credibility of the system and should be avoided as much as possible. That is why it is very important to design and utilize a monitoring and alarm system that (1) will reliably detect and notify the owner of a tank or pipe leak condition and (2) can distinguish between groundwater and leaking chemical product from within the containment system.

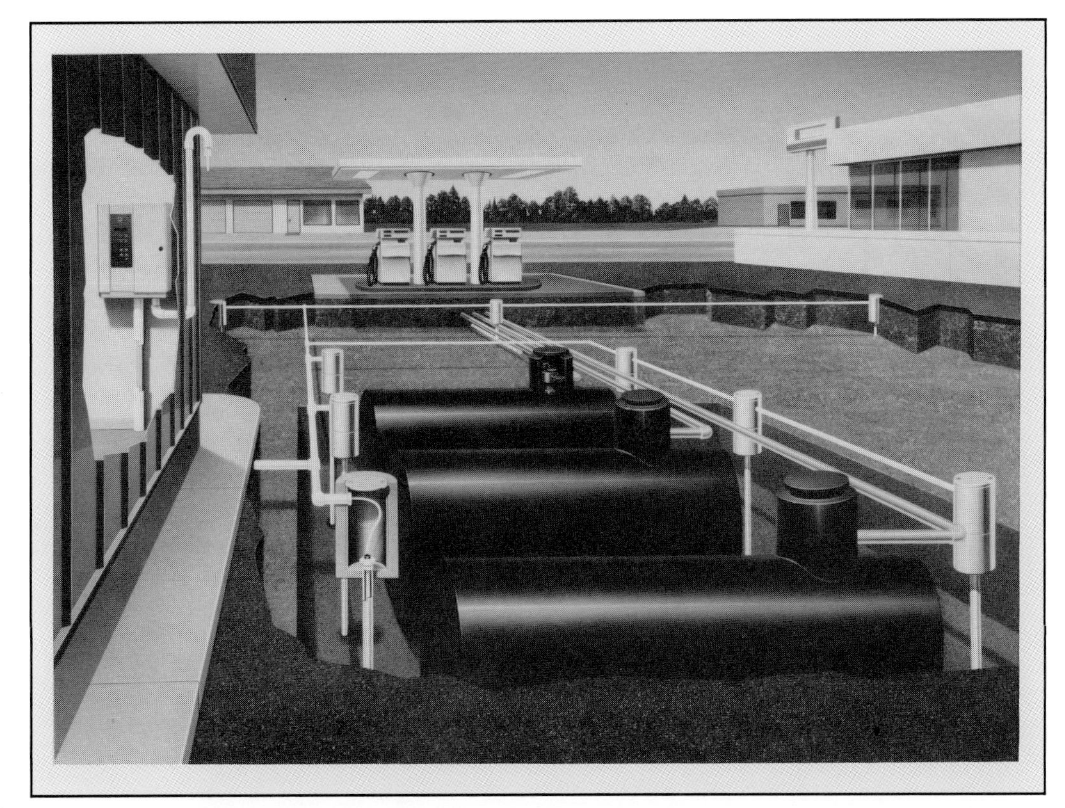

Fig. 5-12. Tank Monitoring System (Courtesy of Arizona Instruments, Inc.)

Underground storage tank (UST) monitoring (see Fig. 5-12) typically requires the installation of multiple sensors around the perimeter of the tanks or piping for detection of liquids other than water (typically petroleum products). Sensors that cannot distinguish between normal groundwater and liquids that might leak from the tanks or piping can result in numerous false leak detection alarms, particularly in areas where the soil surrounding the tank can absorb rain or groundwater.

A variety of specialized sensors have been designed to be effective in various conditions where leak detection is required. These include dry well, wet well, interstitial, vapor, and phase-detection monitoring.

Vapor Detection Process

Aspirated monitoring systems are available that can monitor up to twelve areas in a tank farm backfill for the presence of

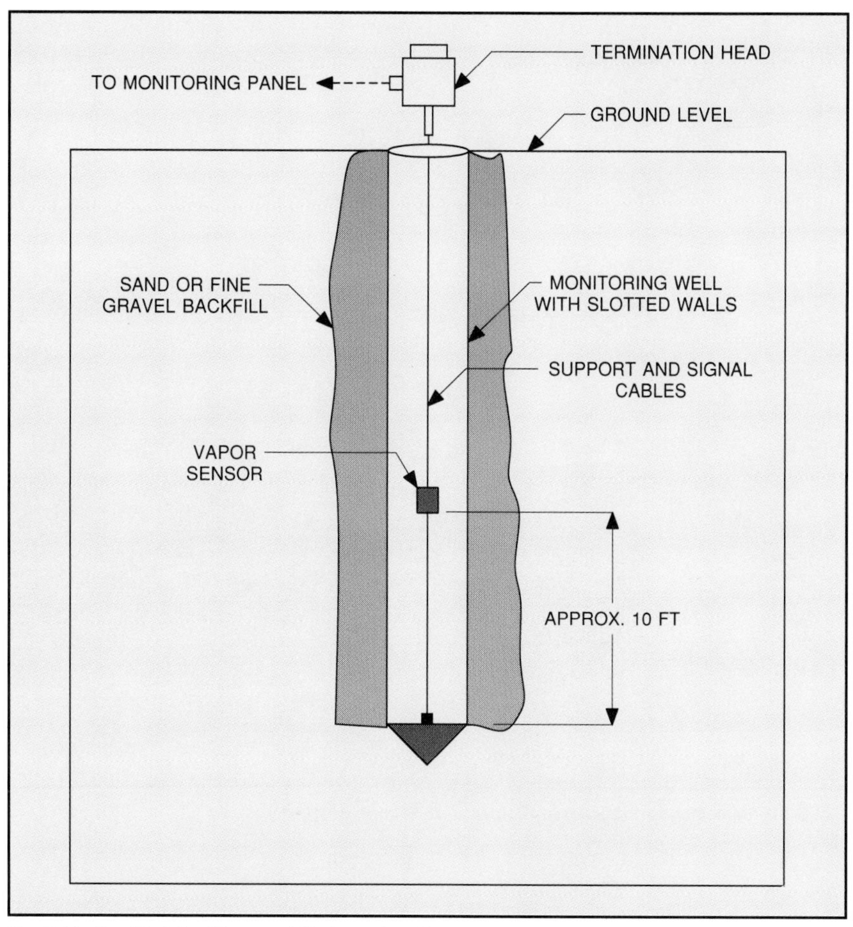

Fig. 5-13. Dry Well for Water and Hydrocarbon Vapor Monitoring

hydrocarbon vapors, indicative of a fuel tank or piping leak (see Fig. 5-13).

It can take days for liquid from minor leaks to permeate surrounding soil and enter nearby monitoring wells, where they are detected on the surface of groundwater. Vapors tend to migrate through soil much more easily and, therefore, faster than liquids—up to 50 times faster, depending upon the soil's permeability. Vapor detection, therefore, often provides earlier detection than liquid detection systems.

An alarm threshold level (in ppm, parts per million) of hydrocarbon vapor is selected and programmed into a microprocessor-based leak detection monitoring panel (see Fig.

5-14). Vapor monitoring wells are strategically located near tanks, pipes, and dispensing stations. The well probes are, in turn, connected to the microprocessor panel's vapor analyzer via polyethylene tubing.

Every eight hours a vacuum pump draws air samples through the analyzer from the vapor monitoring wells, one well at a time. If the detected vapor level at any well exceeds the preprogrammed alarm limit, an audible alarm and an alarm

Fig. 5-14. Tank Monitoring System Microprocessor Panel (Courtesy of Arizona Instruments, Inc.)

indication on the panel face are automatically activated to notify the facility operator of a detected leak and the well sensor that detected it. Date and time of the alarm occurrence are logged for future records.

After each air sample is analyzed, the system automatically purges the analyzer's internal vapor sensor with fresh air and then measures the analyzer's return to a baseline value (reference point). The measured values of samples are stored in memory and/or printed out for record keeping.

Remote communication options are generally available on these systems, allowing monitoring data to be transferred via RS-232 to a local personal computer, remote PC via dial-up modem, or automatic phone dialers preprogrammed with verbal alarm messages in the event of a leak during unmanned hours.

Liquid Detection Process—Tanks

In dry and wet well applications, sensors installed in areas where groundwater is anticipated are designed to distinguish between petroleum products (fuel oils, gasoline) and water. This is accomplished by providing a dual float arrangement at the sensor (see Fig. 5-15).

The first float is constructed so that it is buoyant in water, yet heavier than petroleums. The second float is designed to be bouyant in both water and petroleum. This allows the sensor to float above the water, and, therefore, it will not activate the detection alarm. However, if petroleum is also present (floating in a layer on the water), the sensor becomes wetted by the petroleum and activates the leak alarm. If petroleum is present without groundwater (dry well), it will still be detected by the sensor.

Sensors have also been designed to detect leaks within the interstitial space of a double-wall tank. On this type of storage system (usually a steel tank) a sensor attached to a strap is fished around the interior tank in order to allow it to be positioned at the low point of the outer tank. The strap also allows the sensor to be retrieved easily for inspection (see Figs. 5-16 and 5-17).

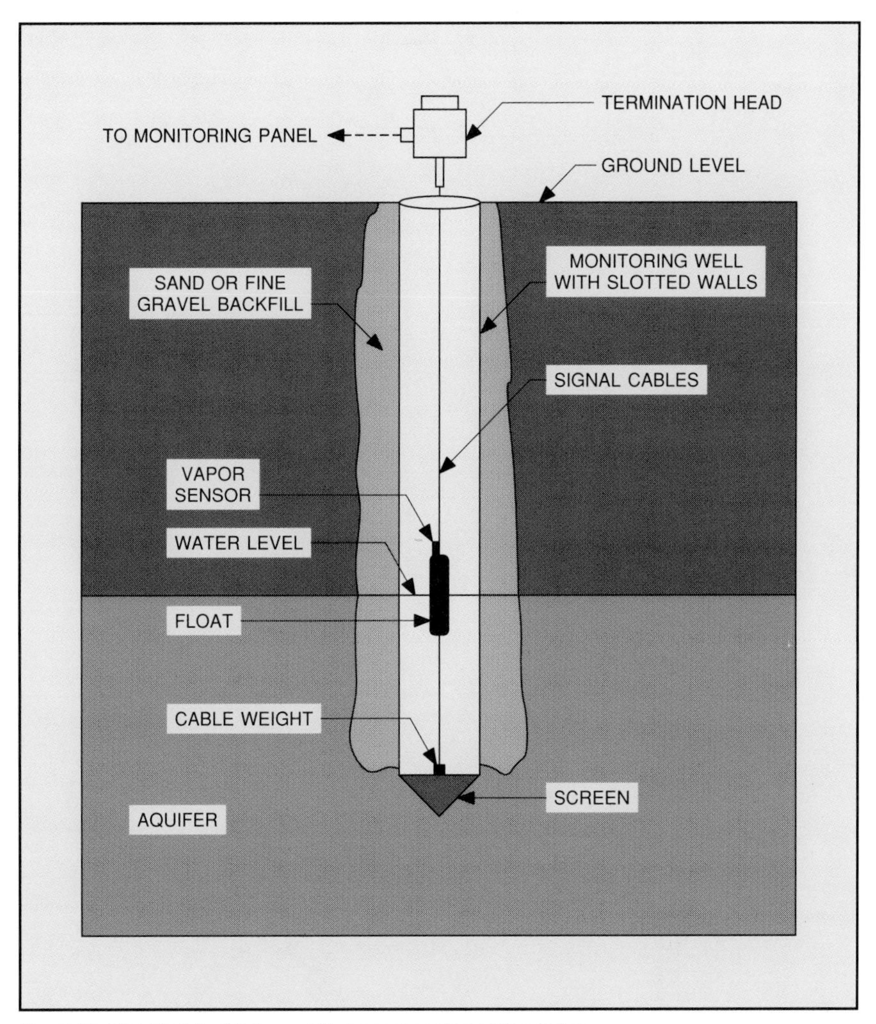

Fig. 5-15. Wet Well for Water and Hydrocarbon Vapor Monitoring

Liquid Detection Process—Piping

On underground piping systems, the secondary containment (outer conduit) of each major pipe section is pitched toward a sump, allowing small quantities of leaking liquid to collect and activate a leak detection sensor located at the bottom of the sump (see Fig. 5-18). Underground piping that is not pitched will require multiple sensors to detect leaks in long sections of pipe, or the bottom of the entire conduit may have to fill with liquid before a leak is detected.

Another leak detection system, designed primarily for double-wall, pressurized piping systems that transport liquids above

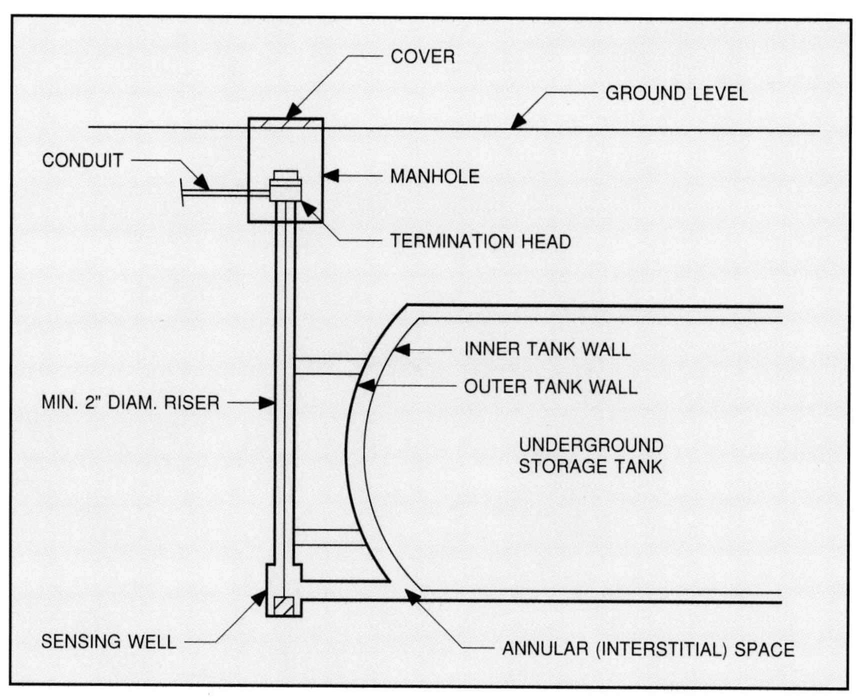

Fig. 5-16. Double-Wall Storage Tank Monitoring

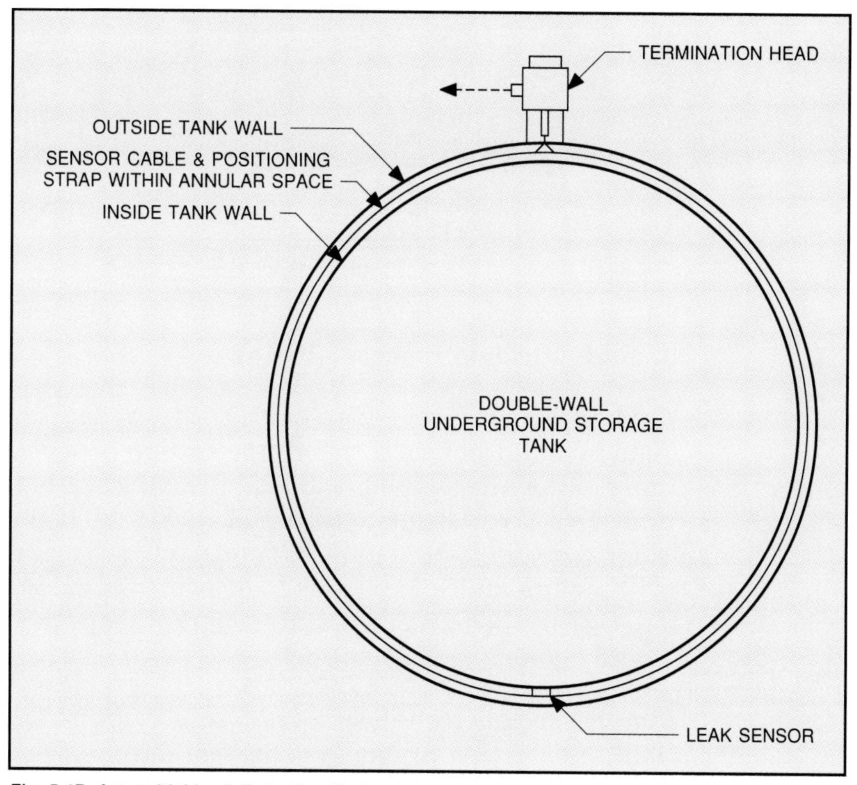

Fig. 5-17. Interstitial Leak Detection System

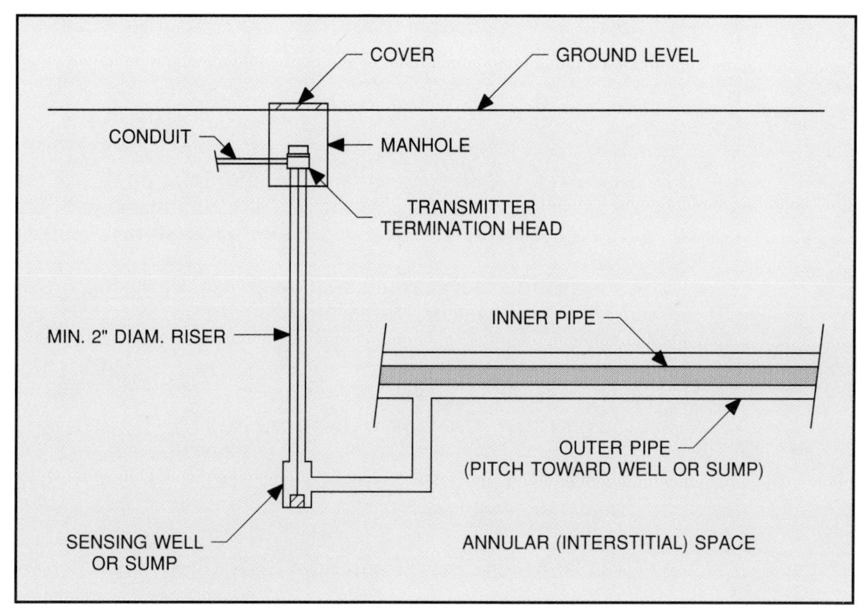

Fig. 5-18. Double-Wall Pipe Leak Detection

ground, uses pressure sensors within each major pipe section and fittings (see Fig. 5-19).

Under normal conditions (no leaks) pressure within the outer pipe wall (conduit) is at atmospheric pressure. This pressure will vary somewhat with changes in ambient air or inside pipe surface temperature due to the expansion of trapped air. If a leak develops within a section of the piping system, pressure from the interior pipe will begin to pressurize the outer pipe,

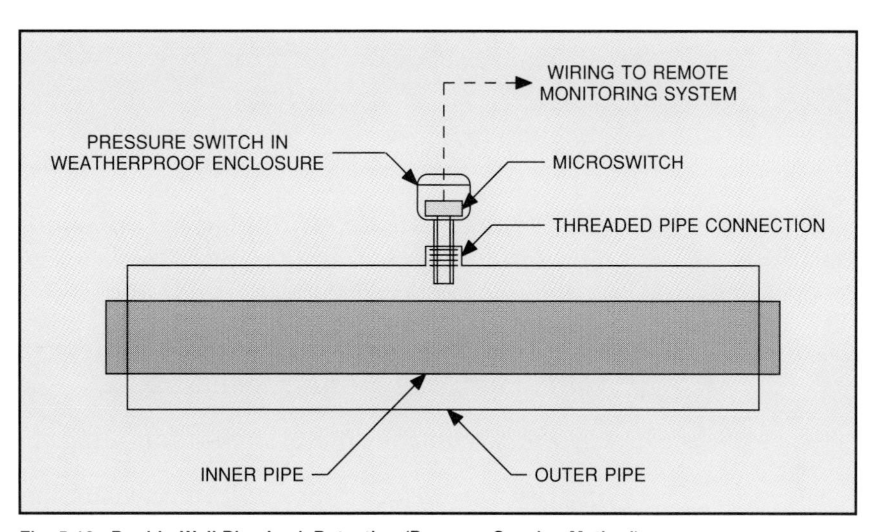

Fig. 5-19. Double-Wall Pipe Leak Detection (Pressure Sensing Method)

causing the pressure switch to actuate and activate a remote alarm.

References

1. Stevens, J. C., *Stevens Water Resources Data Book*, 4th Edition (1988).
2. *Federal Register*, ''A Must for USTs,'' (Summary) U.S. Environmental Protection Agency, Office of Underground Storage Tanks, September, 1988.
3. Comprehensive Environmental Response, Compensation and Liability Act of 1980 (CERCLA or ''Superfund Act'').

Exercises:

5-1. *What are the basic elements of the hydrologic cycle?*

5-2. *Sketch the basic design of a typical groundwater monitoring well.*

5-3. *What factors determine the appropriate location and depth of groundwater monitoring wells?*

5-4. *Design a method of monitoring leaks from the underground storage tank shown in the sketch below.*

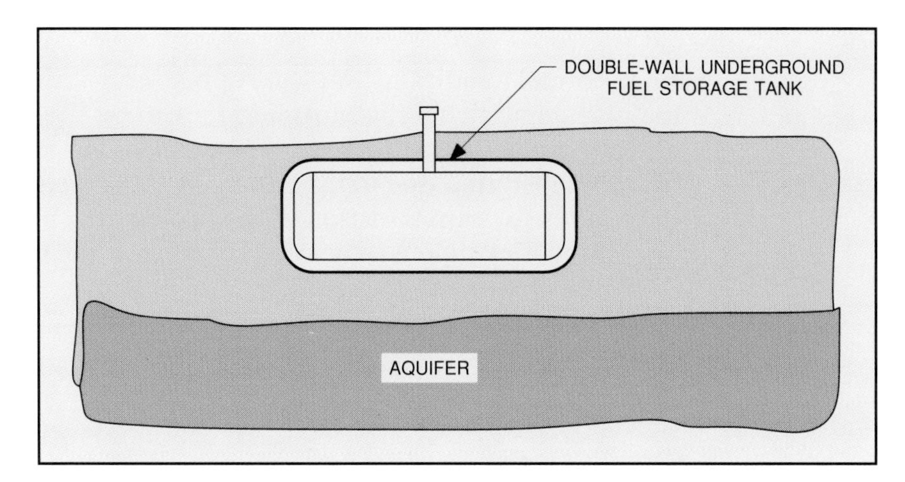

5-5. *What type of leak detection devices are required to monitor double-wall underground fuel storage tank systems installed in a ''wet'' area? In a dry area?*

5-6. *What type of data recording device is preferred for continuous groundwater monitoring?*

5-7. *List the advantages of automatic sampling systems.*

5-8. *Identify the various types of level measurement devices available for determining water level in monitoring wells.*

Unit 6: Soil Decontamination

UNIT 6

Soil Decontamination

Most banks, loan agencies, and developers now require environmental audits of properties before loans are approved for their purchase, or owners are allowed to borrow against the property's value. Significant toxic contamination discovered at the site can greatly reduce property values because of the high cost of decontamination and cleanup.

Large soil decontamination (detoxification) projects can potentially run into hundreds of thousands, perhaps millions, of dollars. This does not include significant (perhaps retroactive) fines that may be levied by the EPA or DEC if significant environmental damage has occurred as a result of blatant violation of government regulations.

Furthermore, sale of the property doesn't necessarily alleviate the former owner of all responsibility for cleanup costs. Legal counseling should be sought in such situations.

Numerous waste sites throughout the U.S. have been designated ''Superfund'' sites because of the millions of tax dollars the federal government has budgeted to assist corporations and municipalities for cleanup of the most severely contaminated areas. Decontamination costs at these sites can be staggering.

Learning Objectives — When you have completed this unit, you should:

A. Be familiar with typical sources of soil contamination.

B. Know what current methods are being used or tested for the detoxification of soil.

C. Know the methods available for soil gas odor control.

D. Be familiar with instrumentation used to analyze soil contamination.

6-1. Soil Analysis

Detection of soil contamination is accomplished by collecting soil or soil vapor samples from the site and having them analyzed at a mobile laboratory, remote laboratory, or by preliminary analysis using a portable analyzer (field screening).

Soil Gas Sampling

To determine the extent of soil contamination and the size of the affected area, a site assessment must be performed. These assessments are performed by extracting and analyzing soil vapor samples. Collection of samples (extraction) is accomplished by drilling a hole in the soil and inserting a sampling line (see Fig. 6-1).

An alternative method is to drive a perforated pipe or tube deep into the soil, attach a sampling (suction) line to the top of the pipe, and draw sample vapors from the top by creating a negative pressure at the pipe's head space.

In both methods, captured vapors are pumped into a vial or bag, sealed, and sent off for analysis.

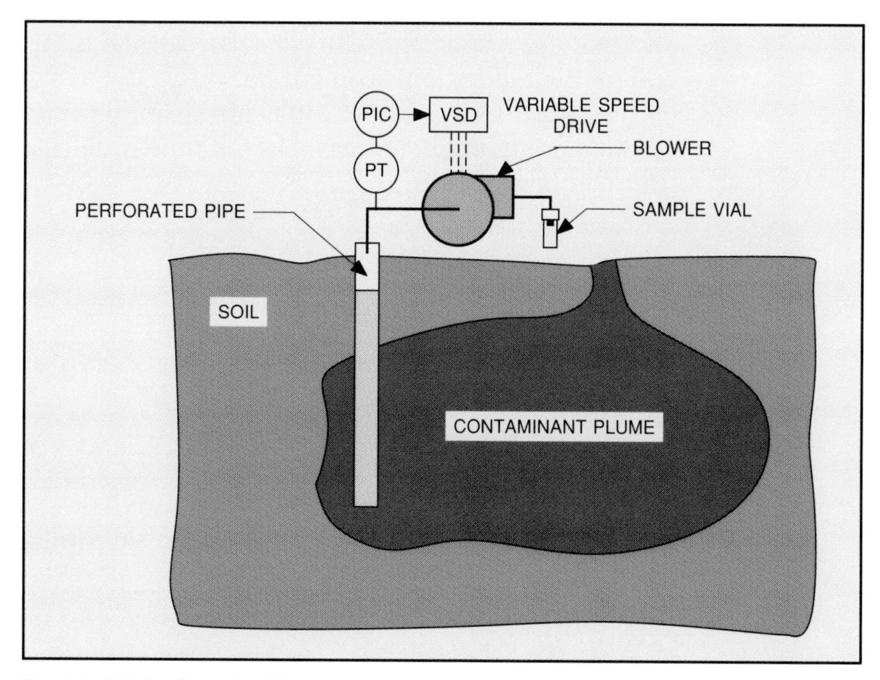

Fig. 6-1. Soil Gas Sampling Line

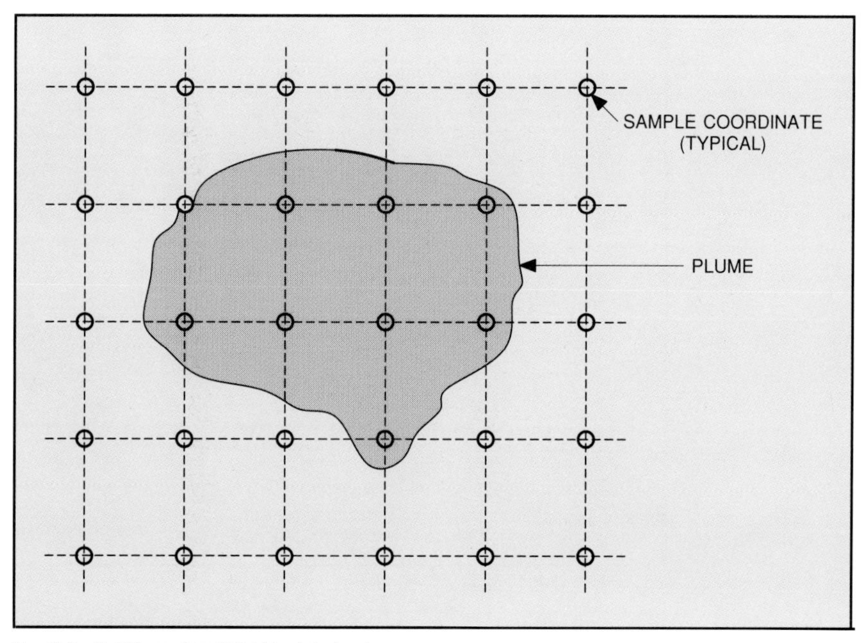

Fig. 6-2. Soil Sampling Grid (Aerial view)

Samples are indexed according to the location from which they were drawn. By establishing a grid pattern of sample points, the area (''plume'') within which contamination exists can be determined (see Fig. 6-2).

Soil samples should also be collected at various depths to determine the maximum depth at which contamination exists and the soil's composition throughout the contaminated soil. This is best accomplished using a hollow stem auger to create a few pilot holes.

If the soil does not contain a lot of rocks, it may be possible to extract soil samples using a sub-soil probe that consists of a 1 to 2-inch hollow tube that is manually driven into the ground using a drop-hammer. The probe is driven to the needed test depth (see Fig. 6-3). When retracted, the soil remains compacted inside the tube. The soil is transported in the tube to the analysis location and then pushed out of the tube into sample vials.

Sophisticated versions of the sub-soil probe have liners that can be removed after extraction and act as a container for transport to the laboratory. This allows the technician to install a new liner and continue sampling with the same probe.

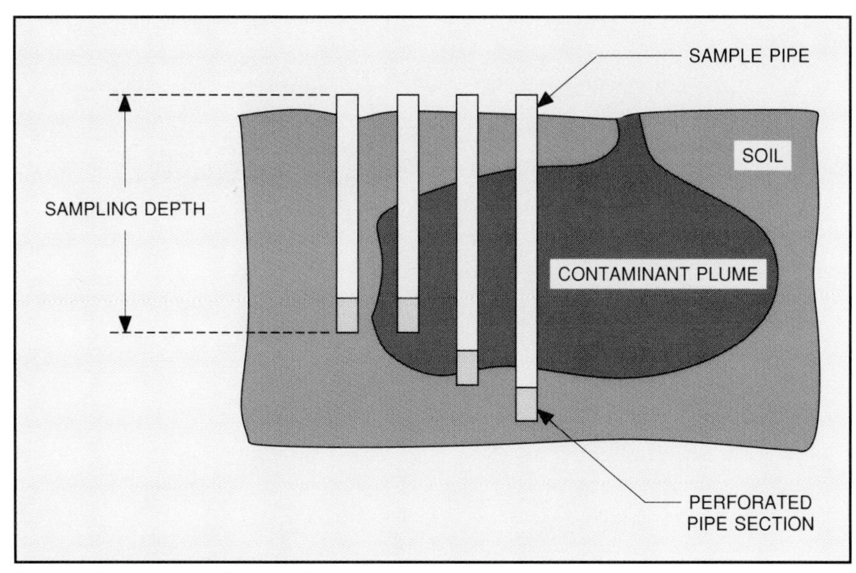

Fig. 6-3. Soil Gas Sampling Depth

Once the vertical and lateral dimensions of the contaminant plume are established and soil composition is known, an appropriate remediation plan can be developed.

Hydrocarbons in Soil

Drawbacks are associated with most methods of soil analysis.

Field screening methods are used to initially determine whether toxic substances are present in the soil. Screening is often accomplished using portable analyzers to perform vapor analysis. Reliability problems can develop when vapor analysis techniques are used as a screening method on weathered samples or samples that are not very volatile to begin with. If a sample is readily volatile, analysis using a portable analyzer can be fast and relatively accurate.

Remote analysis at a laboratory results in a slower response created by transportation time and the analytical laboratory's work load. There is also the potential waste of time and money spent on testing if the samples prove to be negative (uncontaminated) or are lost in transit.

Using analysis by a portable infrared spectrometer, such as the unit in Fig. 6-4, to determine whether volatile or nonvolatile hydrocarbons are present in the soil sample can be performed quickly (approximately 15-20 minutes per sample) in the field.

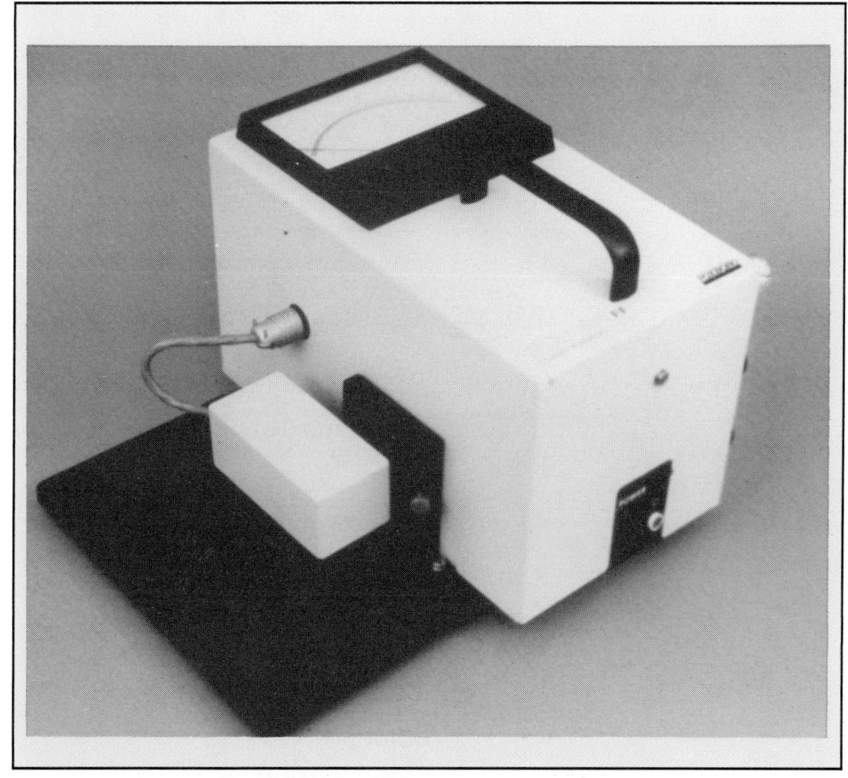

Fig. 6-4. Portable Infrared Spectrometer (Courtesy The Foxboro Company)

Organic vapor analyzers can be used to screen soil samples for the presence of volatile hydrocarbons.

A typical screening analysis using the hydrocarbon vapor analyzer described above would be performed as follows:

A. A 20 gram sample of suspect soil or sand is collected in a reaction vessel. 20 ml of Freon™ are added. The sample and Freon are mixed for 15 seconds, the vessel is then vented, and mixing occurs for another 10 minutes.

B. The extract is slowly decanted through a funnel filter paper and glass wool plug into a container.

C. 20 ml of additional Freon solvent are added to the sample container, mixed for 5 minutes, then decanted again as described above.

D. The funnel system is then washed by pouring 5-10 ml of clean solvent through the filter paper and glass wool. The wash water is then collected in the same container as the sample.

E. Infrared absorbance of the solution is then measured using the analyzer, after it has been allowed 15 minutes of warm-up time (in order to stabilize) and has been zeroed using clean Freon solvent.

F. From a calibration curve, the corresponding ppm of hydrocarbons in the sample is determined.

G. To determine the number of ml of hydrocarbons per kilogram of soil in the test sample, use the conversion equation:

$$\text{ppm} = \frac{\text{mg of hydrocarbon}}{\text{kg of soil}} = \frac{C \times V \times D}{W}$$

where:

C = concentration of hydrocarbons (mg/kg) (from curve)
V = volume of Freon 113 (ml)
D = density of Freon 113 (1.564 g/ml)
W = weight of soil sample (g)

Soil Recovery Rates

Soil and sand have differing recovery rates primarily due to the difference in porosity of the mixture. Therefore, sandy mixtures provide a greater degree of extraction capability than less porous soil or clay.

Examples of sand versus soil recovery rates for several common petroleum products are shown in Table 6-1.

Hydrocarbon	Type	Percent Recovered
Used No.2 fuel oil	Sand	88.4%
Used No.2 fuel oil	Soil	63.7%
Used 10W-40 motor oil	Sand	78.2%
Used 10W-40 motor oil	Soil	74.2%
Weathered gasoline (50%)	Sand	85.8%
Weathered gasoline (50%)	Soil	71.8%

Table 6-1. Hydrocarbon Recovery Data for Sand Versus Soil (Source: The Foxboro Company)

6-2. Soil Remediation Systems

When soil has been contaminated by toxic chemicals, often as a result of leaching, spills, leaks, or condensed chemical vapors, it creates a potential health hazard. Methods employed to decontaminate the area are referred to as remediation processes.

A number of established remedial processes are being used to decontaminate soil. Many more new processes are still under development or awaiting EPA evaluation and approval. The U.S. Environmental Protection Agency must review any new processes to determine their effectiveness in detoxifying the soil without causing any significant damage to the ecosystem in some other manner and to ensure that the general public is not placed at risk.

The most straightforward method of soil remediation simply involves removing and replacing the contaminated layer of soil with fresh uncontaminated earth transported from an uncontaminated area. The contaminated soil is collected using heavy earth-moving equipment. It is then loaded into hazardous waste-hauling trucks and transported to a hazardous waste site for processing or containment.

If the contaminated area is quite large and the contaminated soil layer extends deep into the ground, this becomes a massive, expensive project. In addition, few hazardous waste landfills are still in operation that can accept large volumes of highly toxic material.

In Situ Remediation Processes

Another relatively simple decontamination process involves heating contaminated soil to a high enough temperature that all toxic materials are destroyed. The high temperature used to effectively destroy the contaminants actually fuses the soil into a large block of silicon or glass. This process takes place at the contaminated site. It offers the advantage of leaving the soil in place and intact, thereby limiting worker exposure levels.

Yet another process under development uses a liquid material that is poured into the contaminated soil until the soil becomes saturated. This liquid solidifies after a short period of time, encasing the contaminants. This process, much like the high temperature process described above, prevents contaminants from leaching out of the soil and finding their way into nearby groundwater. Like the previous process, this method allows the soil to be treated in place, limiting worker exposure to the contaminants.

Processes have also been developed and patented by a handful of remediation firms to destroy hazardous dioxins in soil using a biological process. In this process, the contaminated topsoil is exposed to living microorganisms that attack and digest dioxins, rendering them harmless. The soil is then washed and returned to the area from which it was removed. A waste treatment process is set up at the decontamination site. The layer of contaminated soil is collected and fed into the batch process.

Typically, the first stage of the treatment process separates large rocks and debris from the soil by screening and/or a centrifuge. Larger materials are thoroughly washed of any contaminated soil and removed from the process. The wash water is decontaminated.

Remaining soil is screened and then exposed to microorganisms within an environment that encourages their growth and effectiveness. These microorganisms digest dioxins as food. After the process is complete, any residual materials and soil are nontoxic.

Soil Gas Extraction

A method to remove gas vapors trapped within soil involves the construction of multiple dry wells. Vapors from the surrounding soil gradually find their way into the negatively pressurized well through screens in the well walls. These vapors are then drawn out of the wells and through absorption/reaction towers (wet scrubbers) using large blowers. High efficiency absorbent packing materials and chemical sprays efficiently capture chemical vapors present in the air stream. Contaminant removal is accomplished by passing the air through a filter pack of high surface area, nonclogging media, such as polypropylene, that is constantly wetted by the spray nozzles. Fumes are absorbed by the scrubbing liquid, allowing them to impinge upon the packing material and be collected in the wash water.

The remaining relatively clean effluent air is then discharged from the scrubber to atmosphere. Contaminated water from the scrubber process can then be disposed of as hazardous waste.

These air stripping systems are usually custom designed to meet the requirements of a specific application. The number of wells and size and type of packed absorption/reaction tower are factors in sizing and selecting corrosion-proof and sparkless blowers. The blower must provide sufficient capture velocity at all wells to collect all the fumes. At the same time, it must also overcome a large pressure drop introduced by the tower's packing materials.

The scrubber spray is chemically treated to neutralize the pH level using either sodium hypochlorite or sodium hydroxide as a reagent.

Literally hundreds of scrubber system configurations are available for efficient gas removal (see Fig. 6-5). An efficient,

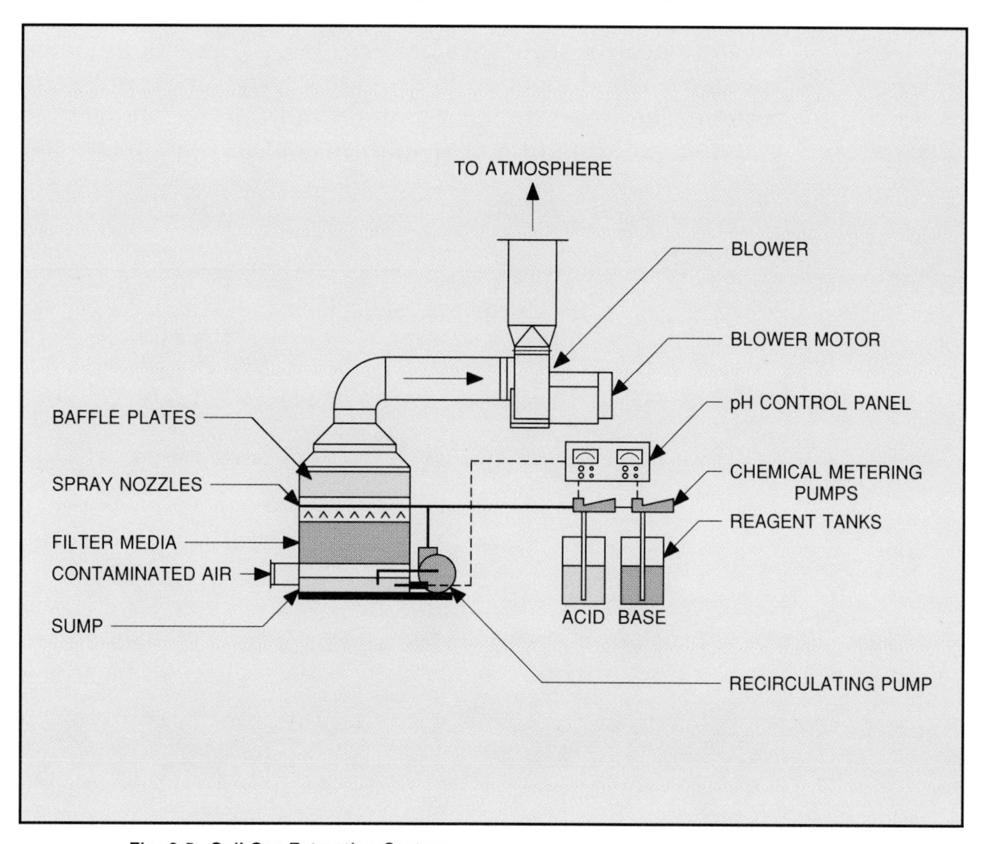

Fig. 6-5. Soil Gas Extraction System

reliable, and cost-effective system should be selected. To ensure good reliability, it is best to keep the system as simple as possible. The more complex the system's operation, the greater the potential for equipment problems and operator error.

6-3. Landfills

Early landfills consisted of large pits into which almost any type of waste material was dumped. Once the pit was full, the waste material was covered by a thin layer of soil. Then a new pit was created and the process begun again. Unfortunately, many toxic materials such as lead, chromium, and mercury were dumped into these waste disposal sites as well. There was little concern about decompositon of the waste and leaching from rainwater, which could eventually allow these chemicals to migrate to groundwater some distance from the dump site.

Newly designed disposal sites have containment systems that consist of a heavy nonbiodegradible, nonpermeable liner system that prevents rain or groundwater from carrying toxic waste to nearby groundwater. Very large sheets of the material are heat welded together to provide a waterproof containment system. The liners are also durable and pliable, allowing them to withstand ground movement caused by climatic changes.

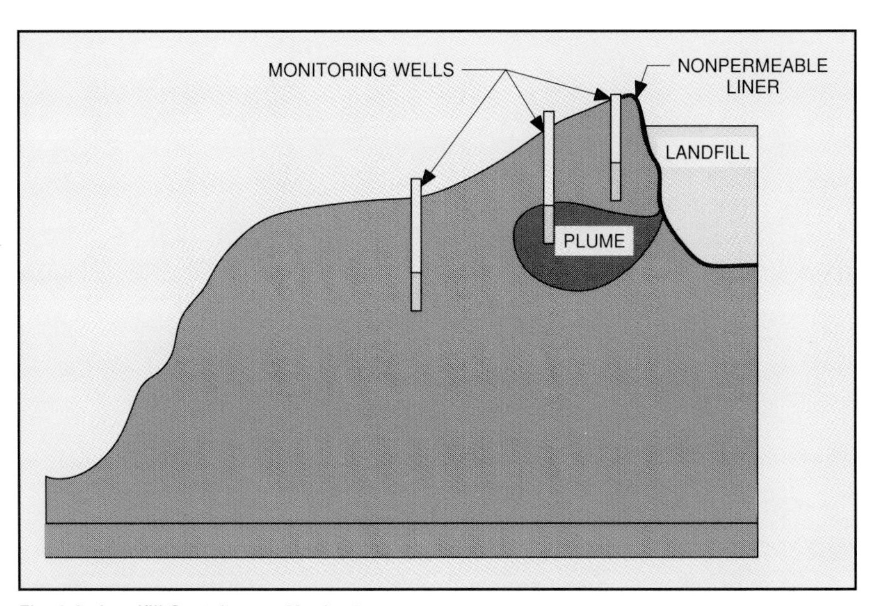

Fig. 6-6. Landfill Containment Monitoring

Since no containment system is totally foolproof, groundwater monitoring wells are established around the perimeter of the dump site to allow continuous monitoring for any leaks (see Fig. 6-6).

References

1. Torpy, M. F., Stroo, H. F., Brubaker, G., ''Biological Treatment of Hazardous Waste,'' *Pollution Engineering*, May, 1989.
2. Johnson, Nancy P., Cosmos, Michael G., ''Thermal Treatment Technologies for Haz Waste Remediation,'' *Pollution Engineering*, October, 1989.
3. Cuddeback, John E., ''Assessments, Audits, and Analysis,'' *Pollution Engineering*, September, 1989.
4. Shelley, Suzanne, ''Turning Up the Heat on Hazardous Waste,'' *Chemical Engineering*, October, 1990.
5. Zumberge, J. E., ''Determination of Semi-volatile Base/Neutral Organic Pollutants in Soils,'' *Pollution Equipment News*, February, 1990.

Exercises:

6-1. *What is the purpose of an environmental audit?*

6-2. *Describe three methods used to perform site audits.*

6-3. *What determines the depth and area of soil to be sampled? What type of analyzer is typically used at the test site to determine whether further analysis is needed?*

6-4. *What soil conditions are generally considered the most difficult to analyze? Why?*

6-5. *How are soil gas samples collected? What methods are used to remove the gases and odors from the exhaust air stream?*

6-6. *Identify and describe several methods of in situ soil remediation now being employed. What are the advantages/disadvantages of in situ remediation?*

Unit 7: Laboratory Analysis

UNIT 7

Laboratory Analysis

This section provides a basic introduction to analytical instrumentation typically used in environmental laboratories for the analysis of air, water, and soil contaminants.

Rapid positive identification and quantification of a toxic substance is often vital for the protection of animal and aquatic life as well as the health and welfare of people. The quicker a potentially hazardous situation is identified, the faster a response plan can be developed and action taken to safeguard people and the environment.

Ideally, one would hope that quick and accurate analysis of contaminants could be performed at the sample site. Often this is very difficult or impractical because the sophisticated laboratory instrumentation required for positive identification of unknown substances is often delicate and requires a controlled operating environment.

A controlled laboratory environment generally means an ambient air temperature somewhere between 60°F and 80°F and relative humidity between 30% and 60% (non-condensing). The environment is relatively dust-free, vibration isolated, and has power line conditioning with surge suppression. This type of environment can be established at the site in a properly equipped portable test laboratory within a trailer. However, this approach can be expensive and often restricts the lab's services to analysis at that particular site for the duration of the project, particularly when the project site is in a very remote location.

New portable analyzers, durable enough to survive field conditions, are being developed for quick on-site analysis of a number of common compounds. A few of these portable or CEM devices were discussed briefly in earlier units of this book and are covered to some extent within this unit.

Learning Objectives — When you have completed this unit, you should:

A. Be familiar with the typical analysis techniques used in analytical laboratories.

B. Understand the basics of how some of the more common analytical instruments work.

C. Know what type of analytical procedure is best suited for analysis of a sample material.

7-1. Qualitative and Quantitative Analysis

For all intents and purposes, any physical property of an element or compound can be used as the basis for analysis. The ability to absorb light, conduct electricity, or conduct heat (in the case of a gas) are fundamental properties commonly used for laboratory and field analysis.

Qualitative analysis determines the presence of a substance within a sample. It provides only a rough idea of the actual quantity of a substance that is present.

Quantitative analysis, on the other hand, provides more precise information about the quantity of a specific sample component.

The physical properties of elements and compounds that have been used for chemical analysis of samples include the following:

A. Mass (or weight)

B. Volume

C. Specific gravity (or density)

D. Viscosity

E. Absorption of radiant energy (infrared, ultraviolet, X-ray, microwaves, and visible light)

F. Turbidity

G. Refractive index

H. Dispersion

I. Nuclear and electron magnetic resonance

J. Electrical conductivity

K. Magnetic susceptibility

L. Heat of reaction

M. Thermal conductivity

N. Radioactivity

Very few analytical methods of analysis are specific to a particular element or compound. Therefore, it is often necessary to perform quantitative separation in order to isolate the desired constituent or remove any interfering substances from the sample before analysis. Some methods for performing such a quantitative separation include the following:

A. Precipitation

B. Distillation

C. Solvent extraction

D. Ion exchange

E. Dialysis

F. Adsorption chromatography

G. Formation of complexes

H. Electrodeposition

Selection of an analytical method depends upon the kind of sample (gas, solution, alloy, etc.), the type of information desired, the required measurement precision, the composition of the sample (number of constituents and amount of desired constituent present), the speed of analysis, the sample destruction or conservation requirements, and the availability and cost of the necessary test apparatus.

The following sections will discuss basic chemical analysis techniques commonly used for laboratory (and, in some cases, field) analysis of a sample substance. Some of the major advantages and disadvantages of each process will also be identified to aid in the selection process.

7-2. Gas Chromatography

One of the oldest and most widely used laboratory instruments for analysis of volatile organic compounds (VOCs) is the gas chromatograph or GC (see Fig. 7-1). Gas chromatography is not a new analytical technique; this method of analysis has been around since the early 1950s.

In the GC analytical process, a sample material is injected into a long, glass capillary column that is filled with a coarse, granular solid. The column is enclosed in a high-temperature oven, which causes the sample material to boil off and produce

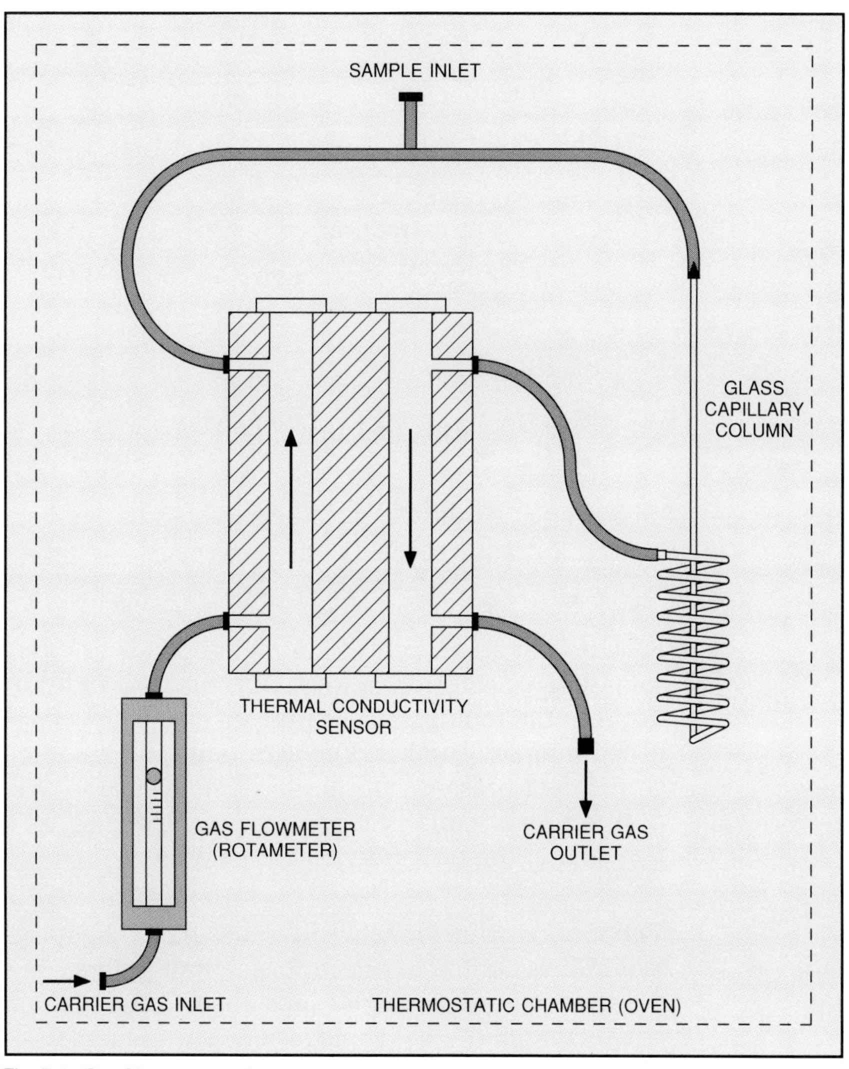

Fig. 7-1. Gas Chromatograph

a vapor. This vapor is then passed through the column, using an inert carrier gas such as helium, nitrogen, or carbon dioxide. Molecular weight and polarity of the various components present in the sample vapor differ somewhat. Therefore, each substance's vapor flows through the column at a slightly different rate. This difference in flow rate causes them to begin to separate within the capillary tube.

Special coatings on the interior wall of the tube contribute to the degree of separation that will occur. Some vapor compounds separate more easily with exposure to some types of coatings than others. Capillary coating selection usually requires a ''trial and error'' approach, particularly when dealing with unknown substances.

It is sometimes difficult to distinguish between two substances whose recorded peak values end up on top of one another. In this situation, another capillary tube may provide better separation of the peaks. Variations in oven temperature can also assist in separating the peak readings. If the peaks cannot be separated by these methods, the compounds are considered to have nearly identical characteristics.

Successful GC analysis requires the sampled material to have a vapor pressure attainable at a temperature that won't destroy the sample or the analyzer.

Very small quantities of a liquid or solid sample (several microliters) are all that is required for GC analysis. Successful GC analysis requires the analyzed material to have a vapor pressure attainable at a temperature that won't destroy the sample or the analyzer.

Once separated, effluent gases from the column are analyzed and measured by a gas detector employing techniques of flame ionization, thermal conductivity, or electronic capture.

Essentially, any physical properties of a sampled gas that are measurably different from other gases and can be easily established could be used as a basis for detection and identification.

Gas chromatography is not well-suited for analysis of polar compounds such as organic acids. Organic acids and similar polar compounds are analyzed more successfully using liquid chromatography (LC). This analytical process is quite similiar to GC. However, in this process, analyzed compounds remain in a liquid state rather than a vapor or a gaseous state.

7-3. Thermal Conductivity

One of the most common methods of gas detection is thermal conductivity. Because of its relatively simple and effective measurement technique, the thermal conductivity analyzer can more easily be adapted for field use. As described in Unit 3, a TCD consists of two metallic blocks, each containing a tubular cavity for the flow of gas. A heated resistance element or thermistor located within the cavity dissipates heat to the block at a rate that is dependent upon the thermal conductivity of the flowing gas.

A "carrier" or reference gas such as helium flows through one of the blocks, and sample gas flows through the other (see Fig. 7-2). Helium and hydrogen have much better heat conducting characteristics than other gases, as seen in Table 7-1. Since hydrogen is highly volatile, helium is the preferred choice for a reference or carrier gas.

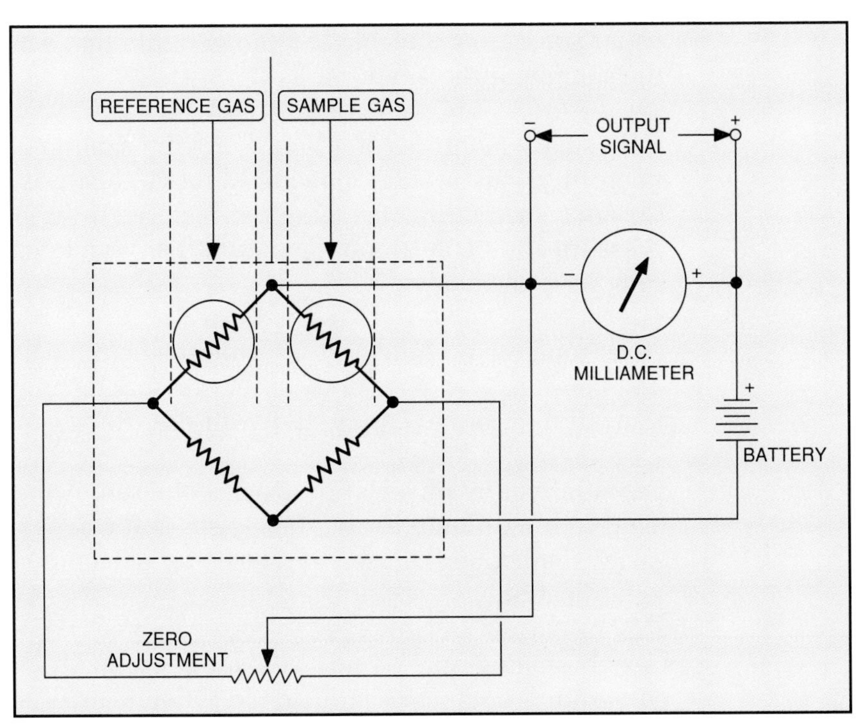

Fig. 7-2. Thermal Conductivity Analyzer

Gas	Thermal Conductivity (kilo-ergs/cm^2/sec @ 0°C)
Air	2.23
Carbon monoxide	2.15
Carbon dioxide	1.37
Chlorine	0.72
Hydrogen	15.90
Helium	13.90
Nitrogen	2.28
Nitrous oxide	1.44
Nitric oxide	2.08
Oxygen	2.33
Sulfur dioxide	0.77

Table 7-1. Thermal Conductivity of Selected Gases and Vapors

Heated elements in each cavity are part of an electrical circuit that forms a Wheatstone bridge. When the reference gas is allowed to flow through both cavities, thermal conductivity of the gas in both cavities is the same. As a result, the heat dissipation at both resistance elements (or thermistors) is equal. Therefore, the Wheatstone bridge circuit is considered "balanced," and the bridge's output signal remains low.

If reference gas in one of the cavities is replaced by a sample gas with different thermal conductivity characteristics, heat dissipation at the sample cavity's resistance element will change, unbalancing the bridge and proportionally changing its output voltage or current.

This bridge output signal (E) is proportional to the partial pressure (p) of a particular vapor present in the carrier gas according to the equation:

$$E = kp$$

where the value k is a proportionality constant equal to the difference in thermal conductivity characterisics of the sample vapor and carrier gas.

Typically, output (E) of the thermal conductivity analyzer is plotted on a graph against (retention) time. The quantity of a particular solute present in the sample is proportional to the area under the recorded peak curve (see Fig. 7-3).

Identification of the peaks created by an unknown in an analyzer's output record can be determined by precalibration or

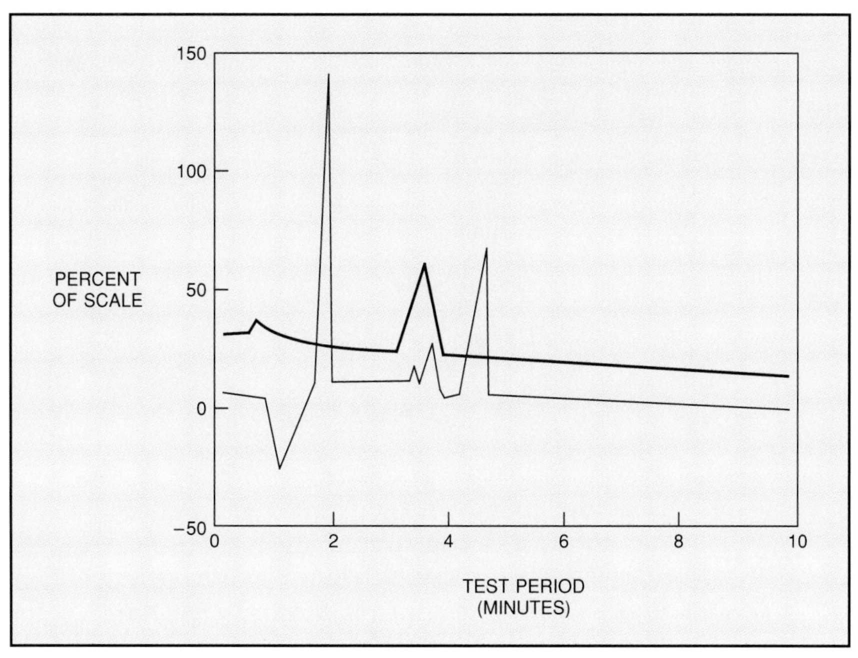

Fig. 7-3. Thermal Conductivity vs. Retention Time

further analysis. In precalibration, the time of appearance (retention) of known gases or vapors is compared to that of the unknown gas to determine the gas that the sample most closely resembles.

If this method is inconclusive, effluent samples are collected from the column after TCD analysis for further analysis using another test method.

Since a thermal conductivity detector does not require any delicate or moving parts, its design is well-suited for rugged, reliable, portable analysis or CEM. Thermal conductivity has been used effectively for years to measure the percentage of carbon dioxide present in flue gas as an indirect indication of combustion efficiency.

7-4. Mass Spectrometry

Mass spectrometry (MS) is a very effective technique for the analysis of unknown compounds. A very small sample of an unknown material is ionized and passed through a magnetic field (see Fig. 7-4). The molecular mass of the sample fragments can then be counted and measured. The fragment pattern is

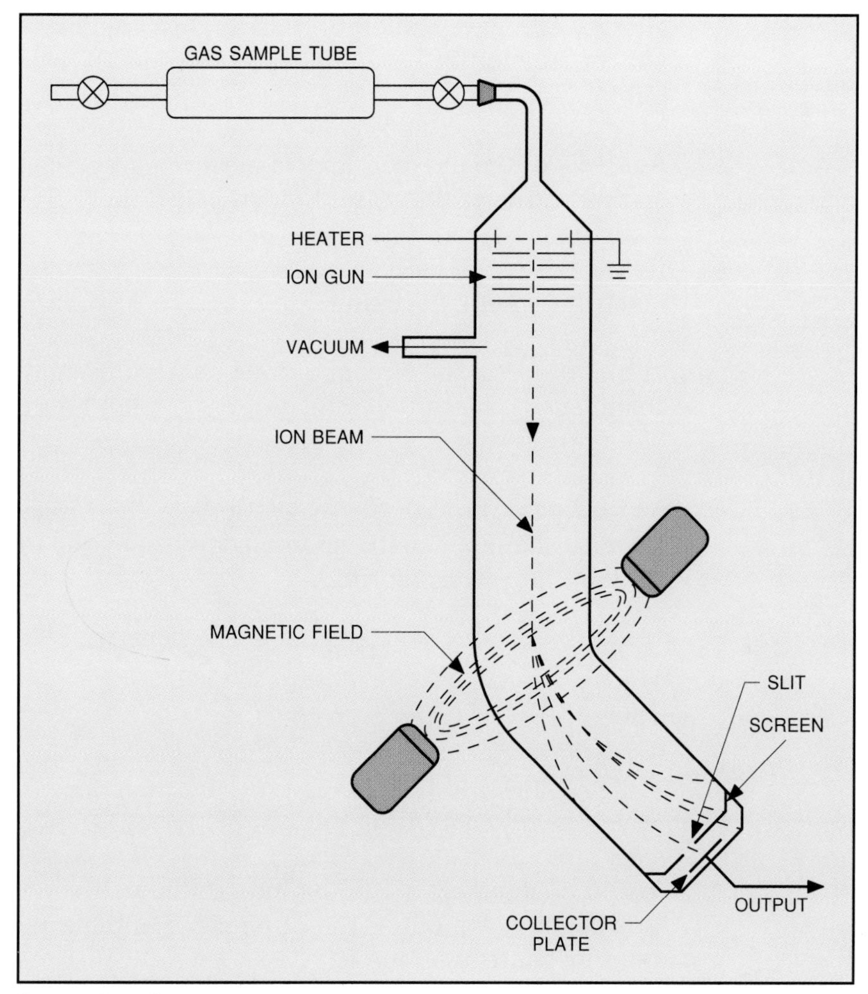

Fig. 7-4. Mass Spectrometer

then recorded and matched with a known structure pattern to provide fairly conclusive identification of the compound.

Mass spectrometers are instruments that sort out charged gas molecules (ions) according to their total mass. Ions are produced by a collision of rapidly moving electrons with molecules of a sample gas being analyzed. This collision causes the electrons to ionize (charge) neutral molecules of the sample gas by knocking their orbiting electrons out of their orbits. The sample molecules then retain a positive charge.

The resulting positive ions are quickly drawn out of the gas stream by a negatively charged accelerating electrode. This removal process works on the basis that objects with opposite charges attract one another. Residual gas molecules that have not become ionized are removed from the MS using a vacuum pump.

Analysis of the charged ions is performed by one of several methods, one of the better known being electromagnetic focusing.

Electromagnetic Focusing

In this process, the stream of ions passes through an evacuated chamber that has a powerful magnetic field, referred to as the "ion gun."

As ions pass through the magnetic field their trajectory and velocity change. Kinetic energy acquired by the ions follows the familiar physics formula:

$$\text{Energy} = 1/2mv^2$$

where:

m = mass of the ion
v = velocity of the ion

As the ions pass through the ion gun in a curved trajectory, they begin to fan out as the differences in their mass are acted upon by centrifugal force.

Analyzers are designed in a manner that allows only those ions with a trajectory matching a predetermined curvature radius (characteristic of the constituent being measured) to reach the ion collector. These ions pass through a very small, adjustable slit in a shield positioned ahead of the collector plate, in relation to the desired constituent's trajectory. Width of the slit is varied to adjust sample resolution. Refer to Figure 7-4.

Time-of-Flight Spectrometry

A second type of mass spectrometer, which uses the time-of-flight method, pulses the ion gun (i.e., acceleration potential) to allow sorting of the ions according to their individual velocities. Since ion velocity varies according to its mass, the time required to get from the ion gun to the collector, or anode, is measured to determine the mass (and ultimately the identity) of the gas constituent. The magnetic field used in the electromagnetic focusing method is not required.

Linear RF Spectrometer

Ions may also be separated by mass using a method referred to as linear radio-frequency spectrometry.

Similiar to the electromagnetic focusing method, electrons are accelerated by a grid to impact molecules and positively ionize them. The ions are then accelerated further by a negatively charged grid. A third grid's charge is alternated at radio frequency between a positive and a negative charge. Ions passing through this grid are either aided or hindered by the charge. The greater the ion's mass, the greater its momentum and the lower the effect of the third grid's alternating charge on its velocity.

Resolution of this type of spectrometer is less than that of electromagnetic focusing and time-of-flight spectrometry, but it is sufficient for analysis of most gases that have a low mass.

Mass spectrometry is not an effective method for the analysis of isomers. Isomers are defined as two or more chemical compounds containing the same elements in the same proportion by weight but differing in molecular structure.

Since analysis by mass spectrometry relies upon a difference in mass to distinguish between two gas constituents, they cannot be distinguished by the MS process.

7-5. Additional Methods of Analysis

Flame Ionization Detection

The flame ionization detector (FID) is a highly selective gas detection system used primarily for the analysis of hydrocarbon gases such as natural gas, propane, butane, ethylene, and low concentrations of chlorinated compounds. It is unaffected by the presence of carbon monoxide or carbon dioxide.

Nuclear Magnetic Resonance

Nuclear magnetic resonance (NMR) is a specialized technique used for the measurement of hydrogen, carbon, and fluorine in an unknown sample. Analysis is performed by subjecting the sample to a strong magnetic field. A distinctive magnetic resonance pattern of the material is displayed on a computer terminal. The pattern indicates whether the target atom is aliphatic (a class of carbon compounds in which carbon atoms are joined in open chains) or aromatic (some derivative of benzene) and its position in relation to other atoms.

Nuclear magnetic resonance is quite effective in the analysis of isomers.

Analysis of Inorganics

Inductively coupled plasma (ICP) is a useful technique for quantitative measurement of almost any inorganic material. In the ICP process, a sample is dissolved, heated into a plasma state, and then passed through a flame. Each atom present in the sample is detected by its unique electronic spectrum.

This type of analyzer is capable of measuring many different elements at the same time. This makes it a rapid analysis method when the sample can be easily dissolved. Unfortunately, some solid samples take hours to dissolve. In such a case, other methods of analysis may be more expedient.

X-Ray Fluorescence

One faster method of analyzing slow-dissolving solids is the X-ray fluorescence (XRF) technique. In this process, an unknown solid or liquid sample is exposed to X-rays emitted by a tungsten filament, causing it to ''fluoresce'' at wavelengths characteristic of a specific element. The characteristic reflected radiation from the sample element is superimposed on a broad spectrum of X-rays.

In addition to sample fluorescence, some X-rays are also scattered back. The intensity of this incidental radiation can be measured to establish the sample's mass and density.

X-ray fluorescence is generally considered effective for identifying and quantifying any element with a molecular weight greater than that of magnesium (molecular weight = 24.312).

X-Ray Diffraction

Another analytical process involving exposure of an unknown sample to X-rays is commonly known as X-ray diffraction (XRD). This technique uses characteristic patterns created by crystalline materials to identify the element. It is effective in identifying an inorganic material but is relatively ineffective in quantifying elements present in the sample (qualitative analysis).

Infrared Spectrometry

Nearly all chemical compounds have a distinct selective absorption level of infrared radiation. See Fig. 7-5 and refer to the chart in Appendix D.

Wavelengths of light absorbed by the constituents of the gas stream correspond to the various gases (see Table 7-2).

Atomic Absorption Spectrophotometry

Atomic absorption spectrophotometry is specified by the EPA as the method to test for lead, arsenic, and selenium.

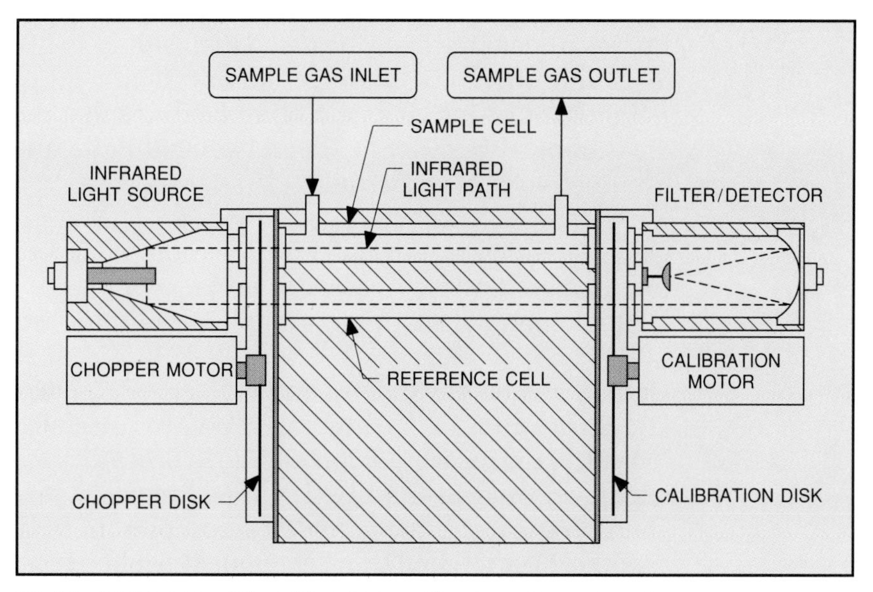

Fig. 7-5. Nondispersive Infrared Gas Analyzer (Courtesy of Anarad, Inc.)

Compound	Wavelength (in micrometers)
Octane	3.4
Carbon dioxide	4.7
Nitric oxide	5.3
Sulfur dioxide	8.8
Methylene chloride	13.5

Table 7-2. Analytical Wavelengths of Common Compounds

Ion Chromatography

Ion chromatography is a method of analysis (EPA Method 300.0) for inorganic anions in water.

Trace organic contaminents that cannot be successfully analyzed using a gas chromatograph have traditionally been identified using high performance liquid chromatography (HPLC). Even this method may be unsuccessful when dealing with polar and ionic organics. In such cases, ion chromatography may be an effective method of analysis.

Fixed-Filter Analyzers

A fixed-filter laboratory analyzer consists of a single-beam infrared spectrometer used primarily to perform analysis of hydrocarbons present in water. This method is EPA-approved for monitoring oil and grease present in effluent water from refineries, drilling rigs, and other facilities that require this measurement in order to obtain National Pollutant Discharge Elimination System permits.

Other facilities requiring NPDES permits include textile mills that process raw wool; seafood processing plants; and manufacturers of soap, tar, asphalt, aluminum, iron, and steel.

Organic matter is extracted from water using Freon 113™ (1,1,2 trichloro-1,2,2 trifluoroethane). Infrared absorbance of the extract is measured at the C-H (hydrocarbon bond) band maximum, and Beer's Law is applied to calculate the concentration of organic hydrocarbon present in the water sample. This method of measurement is effective over a wide range of hydrocarbon concentrations in water.

The same measurement technique is also used for monitoring waste solvent concentrations in water, determining the amount of coatings applied to synthetic fibers during finishing operations, and determining the concentration levels of organic additives to an aqueous bath.

Some analyzers can also be equipped with a battery/inverter option for portable measurements in the field.

Concentrations can be determined in either parts per million (ppm) or milligrams per liter. Levels of hydrocarbons in water can be detected down to 1 ppm using a 1 cm^2 quartz curette as a transmission cell.

Both volatile and nonvolatile oily material can be analyzed per EPA Method 413.2.

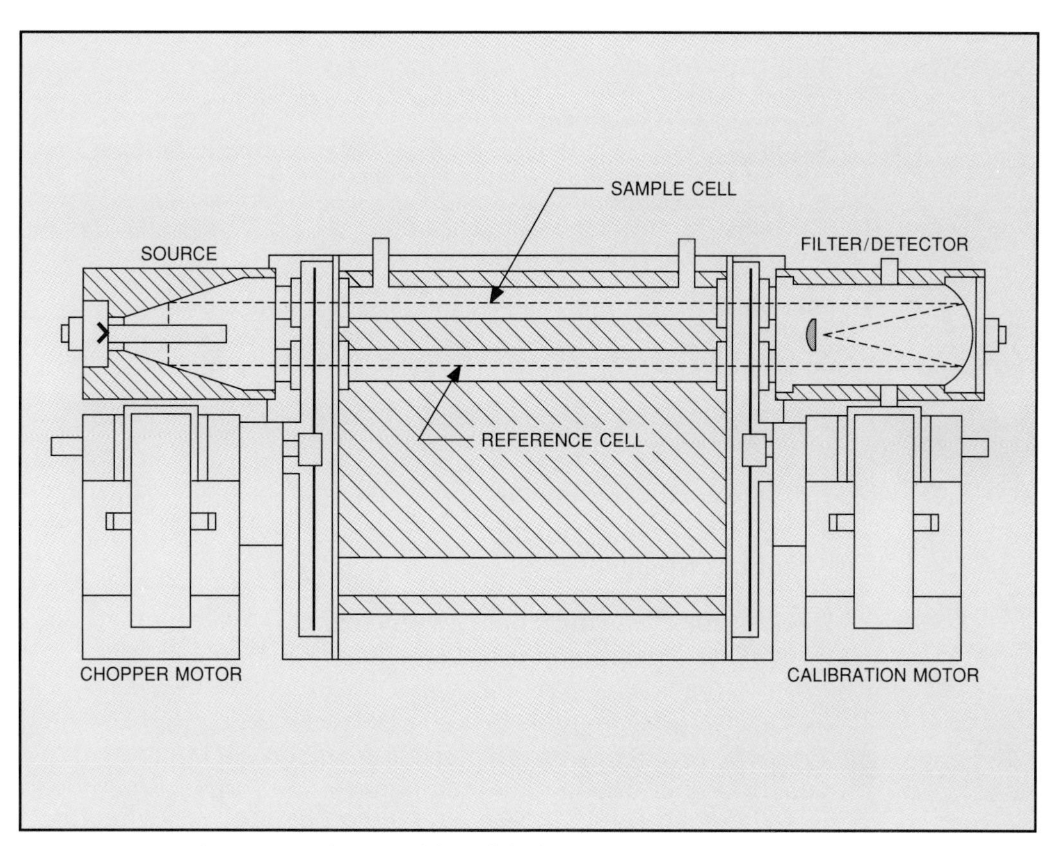

Fig. 7-6. Ultraviolet Gas Analyzer (Courtesy of Anarad, Inc.)

Ultraviolet Analysis

Ultraviolet analyzers operate on the same principle as IR (infrared) analyzers. The wavelengths of light absorbed by a sample material are used to establish the sample's identity. Analysis is performed using a light beam in the "near ultraviolet" range of 200 to 380 nanometers.

Construction of the UV analyzer is also very similar to that of IR analyzers (see Fig. 7-6).

In general, UV analyzers can be used successfully for the analysis of aromatics, carbonyls, and most inorganic salts with an accuracy of ±2% of full scale.

Straight chain hydrocarbons, inorganic gases, and the lower alcohols cannot be accurately identified by this method of analysis.

References

1. Ewing, Galen W., *Instrumental Methods of Chemical Analysis*, McGraw-Hill Book Company (1980).
2. Tyson, George K., and Willis, Reid S., "Successful Online Chemical Analysis," *Chemical Engineering*, December 1990.
3. Joyce, R. J., and Schein, A., "Ion Chromatography: A Powerful Analytical Technique for Environmental Laboratories," *American Environmental Laboratory*, November 1989.
4. Brown, M., "Analysis of Unknown Pollutants in Groundwater and Hazardous Waste Using Liquid Chromatography/Particle Beam Mass Spectrometry," *Proceedings of the Fifth Annual Waste Testing and Quality Assurance Symposium*, II-20 through II-41.

Exercises:

7-1. *What is the appropriate type of analytical instrument for identification of organics? Inorganics? Slow-dissolving solids? Isomers?*

7-2. *Describe the technique used in each of the following types of mass spectrometers to identify an unknown:*
- *Electromagnetic focusing*
- *Time-of-flight*
- *Linear radio frequency*

7-3. *What is an isomer and what method of analysis cannot be used successfully to identify it?*

7-4. *Briefly describe how thermal conductivity is used for identification of a gas or vapor.*

7-5. *What methods can be used to distinguish between two similiar unknowns whose peak chromatograph outputs overlap one another?*

7-6. *What is infrared absorption and how is it used to identify unknowns?*

Appendix A: Suggested Readings and Study Materials

APPENDIX A

Suggested Readings and Study Materials

Independent Learning Modules:

One of your best sources of material for further reading and study of process control and instrumentation are the ILMs published by ISA. They are custom designed and created for this exact purpose. Place a *standing purchase order* to receive new ILMs as they are published.

Handbooks and Manuals:

Liptak, Bela G., *Instrument Engineer's Handbook* (Chilton Book Co., 1970)

Considine, Douglas M. (ed.), *Handbook of Applied Instrumentation* (McGraw-Hill Book Co., 1964)

Textbooks (selected titles):

Andrew, W. G., and Williams, H. B., Applied *Instrumentation in the Process Industries*, 2nd ed. (Gulf Publishing Co., 1980)

Dukelow, Sam G., *Control of Boilers*, 2nd ed. (ISA, 1991)

McMillan, Gregory K., *pH Control* (ISA, 1985)

Murrill, Paul W., *Application Concepts of Process* Control (ISA, 1988)

Technical Magazines and Journals (selected titles):

- *Chemical Engineering*, published by McGraw-Hill, Inc.
- *INTECH*, published by the Instrument Society of America.
- *Plant Engineering*, published by Cahners Publishing Company.
- *Pollution Engineering*, published by Pudvan Publishing Company.

Appendix B: Glossary of Terms and Abbreviations

APPENDIX B

Glossary of Terms and Abbreviations

absorption—Process of one substance entering into the inner structure of another.

adsorption—Adhesion of a thin film of material to the surface of a solid substance.

agar—Gelatinous colloidal extract from a red algae.

AIHA—American Industrial Hygiene Association.

ambient air—Air that surrounds an object.

aquifer—Zone of gravel, sand, or porous rock below the water level containing a large volume of groundwater.

ash—In fuel, inorganic material that is non-combustible.

ASHRAE—American Society of Heating, Refrigeration, and Air Conditioning Engineers.

BACT—Best available control technology.

BOD—Biochemical oxygen demand.

CEM—Continuous emissions monitoring.

CFCs—Chlorofluorocarbons.

cfm—Cubic feet per minute.

CFR—Code of Federal Regulations.

CGA—Cylinder gas audit.

CO—Carbon monoxide; a product of the incomplete combustion of carbon.

COD—Chemical oxygen demand.

combustion efficiency—Ratio of fuel fed to a burner versus the amount of fuel actually burned.

concentration—Quantity of one constituent dispersed within a defined amount of another.

DAS—Data acquisition system.

DCS—Distributed control systems.

dust collector—Device used to capture fly ash or dust from an exhaust gas stream.

EGR—Emission gas recycling.

electrostatic precipitator—Device to remove particulates from an exhaust gas stream by using electrodes to ionize the particles so they can be attracted to, and collected by, oppositely charged metal plates or electrodes.

EMAP—Environmental monitoring and assessment program.

entrainment—Collecting or transporting of particles within a fluid flowing at a high velocity.

EPA—U.S. Environmental Protection Agency.

ESP—Electrostatic precipitator.

fabric filter—Used to capture fly ash and dust in an exhaust gas stream using a fabric-like filter material.

FGD—Flue gas desulfurization.

FGR—Flue gas recirculation.

FIA—Flow injection analysis.

flue gases—Products of combustion.

fly ash—Ash carried by exhaust gases out of a furnace.

FTIR—Fourier transform infrared.

fugitive emissions—Emissions that could not reasonably pass through a stack, chimney or vent.

GC—Gas chromatography.

gpm—Gallons per minute.

HFC—Hydrofluorocarbon.

IAQ—Indoor air quality.

infiltration—Air leakage inward through cracks and interstices, ceilings, walls, or floors of a building.

inorganic—Material that is neither plant nor animal.

LAER—Lowest achievable emission rate.

LEA—Low excess air.

LEL—Lower explosive limit.

LC—Liquid chromatography.

MACT—Maximum achievable control technology.

micron—One thousandth of an inch.

microorganism—Organism of microscopic size, including bacterium, fungus, and protozoan.

MS—Mass spectrometry.

Mw—Megawatt; equal to one million watts of electrical energy.

NAAQS—National Ambient Air Quality Standards.

natural draft—Draft caused solely by the effect of the exhaust stack.

NESHAP—National Emission Standards for Hazardous Air Pollutants.

NIOSH—National Institute of Occupational Safety and Health.

NOx—Nitrogen oxides; Unspecified oxides of nitrogen.

NPDES—National Pollutant Discharge Elimination System.

NSPS—New Source Performance Standard.

ODP—Ozone depletion level.

organic—Material composed of plant or animal matter.

OSHA—Occupational Safety and Health Association.

particulate matter—State in which a solid or liquid substance exists in the form of aggregate molecules or particles ranging between 0.01 and 100 micrometers.

PCB—Polychlorinated biphenyls.

PEL—Permissable exposure level.

pH—An index of acidity or alkalinity.

PIC—Product of incomplete combustion.

PID—Proportional-plus-integral-plus-derivative control.

plenum—Air passageway (ductwork).

ppm—Parts per million.

PSD—Prevention of significant deterioration.

psi—Pounds per square inch.

purge—Forcing of clean gas (such as air or nitrogen) through a pipe or chamber to remove unwanted residual gases or vapor.

RAA—Relative accuracy audit.

RATA—Relative accuracy test audit.

RCRA—Resource Conservation and Recovery Act.

respirable—Capable of being inhaled.

S—Chemical symbol for sulfur.

SAMA—Scientific Apparatus Makers Association.

SARA—Superfund Amendments and Reauthorization Act.

SCR—Selective catalytic reduction.

scrubber—Device used to remove particulate matter from an exhaust gas stream; may be wet or dry process.

set point—An input variable that sets the desired value of the controlled variable.

SGR—Soil gas recovery.

skimmer—A surface blowdown device used to remove suspended solids from the surface of a liquid.

SNCR—Selective non-catalytic reduction.

soda ash—A substance such as sodium carbonate that is used to treat wastewater.

SOx—Sulfur oxides; Undefined oxides of sulfur such as sulfur dioxide (SO_2) or sulfur trioxide (SO_3).

stack—A flue, vent, chimney, or other passage through which smoke or hot gases are released to atmosphere.

stack effect—The movement of hot gases out of a stack due to the difference in density between the flue gas and surrounding (cooler) atmosphere.

stratification—The separation of fluids into layers.

suspended solids—Solids not dissolved in water but held in suspension.

TC—Thermal chromatography.

TCE—Trichloroethylene.

TDS—Total dissolved solids.

TOV—Total organic vapor.

toxic—Of, related to, or caused by, a poison.

TSCA—Toxic Substance Control Act.

turbidity—The optical obstruction or deflection of a ray of light through a sample of water and caused by finely suspended matter.

VAV—Variable air volume.

VOC—Volatile organic compound.

Appendix C: EPA Regional Offices

APPENDIX C
EPA Regional Offices

Region I—Connecticut, Maine, Massachusetts, New Hampshire, Rhode Island, Vermont

Regional Administrator........617/565-3400
Deputy Regional Administrator........617/565-3402
Director, Air Management Division........617/565-3800
Chief, State Air Programs Branch........617/565-3245
Acting Chief Technical Support Branch........617/565-3221
Director, Waste Management Division........617/573-5700
Chief, Control Technology & Compliance........617/565-3258
Director, Office of Public Affairs........617/565-3424
Address: John F. Kennedy Federal Building
Boston, Massachusetts 02203
Director, Environmental Services Division........617/860-4315
Address: 60 Westview Street
Lexington, Massachusetts 02173

Region II—New Jersey, New York, Puerto Rico, Virgin Islands

Regional Administrator........212/264-2525
Acting Regional Administrator........212/264-2525
Director, Air & Waste Management Division........212/264-2301
Chief, State Air Programs Branch........212/264-2517
Chief, Air Compliance Branch........212/264-9627
Director, Office of External Affairs........212/264-2515
Address: 26 Federal Plaza
New York, New York 10278
Director, Environmental Services Division........202/321-6754
Address: Woodbridge Avenue
Edison, New Jersey 08817

Region III—Delaware, District of Columbia, Maryland, Pennsylvania, West Virginia, Virginia

Regional Administrator........215/597-9814
Deputy Regional Administrator........215/597-9812
Director, Hazardous Waste Management Division........215/597-8131
Director, Air Management Division........215/597-9390
Chief, State Air Programs Branch........215/597-9075
Air Enforcement Branch........215/597-3989
Director, Office of Public Affairs........215/597-9370
Director, Environmental Services Division........215/597-4532
Address: 841 Chestnut Street
Philadelphia, Pennsylvania 19107

SOURCE: *Journal of the Air and Waste Management Association*

Region IV—Alabama, Florida, Georgia, Kentucky, Mississippi, N. Carolina, S. Carolina, Tennessee

Regional Administrator....404/347-4727
Deputy Regional Administrator....404/347-4727
Director, Air, Pesticides, and Toxics Management Division....404/347-3043
Chief, State Air Programs Branch....404/347-2864
Director, Waste Management Division....404/347-3454
Chief, Air Compliance Branch....404/347-2904
Director, Office of Public Affairs....404/881-3004
Director, Office of External Affairs....404/347-3004
Address: 345 Courtland Street, N.E.
Atlanta, Georgia 30365
Director, Environmental Services Division....404/546-3136
Address: College Station Road
Athens, Georgia 30613

Region V—Illinois, Indiana, Michigan, Minnesota, Ohio, Wisconsin

Regional Administrator....312/353-2000
Deputy Regional Administrator....312/353-2000
Director, Air Management Division....312/353-2213
Chief, Air and Radiation Branch....312/353-2211
Director, Waste Management Division....312/886-7579
Chief, Air Compliance Branch....312/353-2081
Director, Office of Public Affairs....312/353-2072
Address: 230 South Dearborn Street
Chicago, Illinois 60604
Director, Environmental Services Division....312/353-3808
Address: 536 South Clark Street
Chicago, Illinois 60605

Region VI—Arkansas, Louisiana, New Mexico, Oklahoma, Texas

Regional Administrator....214/655-2100
Deputy Regional Administrator....214/655-2100
Director, Hazardous Waste Management Division....214/655-6700
Chief, Air Branch....214/655-7204
Director, Air, Pesticides, and Toxics Division....214/655-7200
Air Compliance Contact....214/655-7220
Director, Office of Public Affairs....214/655-2200
Director, Environmental Services....214/655-2210
Address: 1445 Ross Avenue, 12th Floor
Dallas, Texas 75270

Region VII—Iowa, Kansas, Missouri, Nebraska

Regional Administrator....913/551-7000
Deputy Regional Administrator....913/551-7000
Director, Waste Management Division....913/551-7050
Chief, Air Branch....913/551-7020
Director, Air and Toxics Management Division....913/551-7020
Air Compliance Contact....913/551-7020
Director, Office of Public Affairs....913/551-7003

Director, Environmental Services....................913/551-7042
Address: 726 Minnesota Avenue
Kansas City, Kansas 66101

Region VIII—Colorado, Montana, N. Dakota, S. Dakota, Utah, Wyoming

Regional Administrator....................303/293-1603
Deputy Regional Administrator....................303/293-1603
Director, Waste Management Division....................303/293-1720
Chief, Air Programs Branch....................303/293-1753
Director, Air and Toxics Division....................303/293-1438
Air Compliance Contact....................303/293-1610
Director, Office of External Affairs....................303/293-1692
Address: 726 Minnesota Avenue
Kansas City, Kansas 66101
Director, Environmental Services Division....................303/236-5061
Address: Denver Federal Center
Lakewood, Colorado

Region IX—Arizona, California, Hawaii, Nevada, Guam, American Samoa

Regional Administrator....................415/556-6478
Deputy Regional Administrator....................415/556-6478
Director, Toxics and Waste Management Division....................415/744-1397
Chief, Air Programs Branch....................415/556-5262
Director, Air Division....................415/556-6404
Air Operations Branch....................415/556-6025
Director, Office of External Affairs....................415/556-6383
Director, Office of Policy and Management....................415/556-6458
Director, Environmental Services Branch....................415/556-5043
Address: 215 Fremont Street
San Francisco, California 94105

Region X—Washington, Oregon, Idaho, Alaska

Regional Administrator....................206/442-1220
Deputy Regional Administrator....................206/442-5810
Director, Air and Toxics Division....................206/442-4152
Chief, Air Programs Branch....................206/442-1275
Director, Waste Management Division....................206/442-1352
Air Compliance Contact....................206/442-8507
Director, Office of Public Affairs....................206/442-1466
Director, Environmental Services Division....................206/442-1295
Address: 1200 Sixth Avenue
Seattle, Washington 98101

Appendix D: OSHA Concentration Limits for Gases and Vapors

(This listing of the OSHA Concentration Limits was provided by The Foxboro Company)

Compound	OSHA Limit (ppm)	1990 TWA	1990 STEL	Ceiling	Wavelength (Microns)	Pathlength (Meters)	Absorbance (AU)	Linear Term (ppm/AU)	Min DL (ppm)
Acetaldehyde	200	100.00	150.0		9.3	20.25	0.233	864.2	ERR
Acetic Acid	10	10.00			8.6	20.25	0.072	167.6	ERR
Acetic Anhydride	5			5.0	8.9	20.25	0.180	27.8	0.05
Acetone	1000	750.00	1000.0		8.5	2.25	0.396	2755.1	ERR
Acetonitrile	40	40.00	60.0		9.7	20.25	0.007	6746.3	ERR
2-Acetylaminofluorine	b								
Acetylene dichloride, see 1,2 dichloroethylene									
Acetylene tetrabromide	1	1.00			9.0	20.25	0.001	1877.8	1.2
Acetylsalicylic acid (aspirin)	b								
Acrolein	0.1 a	0.10	0.3		8.6	20.25		476.2	0.9
Acrylamide—Skin	b								
Acrylic acid (Skin)		10.00			8.9	20.25	0.346	295.2	
Acrylonitrile—Skin	2				10.6	20.25	0.005	457.7	0.6
Aldrin—Skin	b	1.00							
Allyl Alcohol—Skin	2	2.00	4.0		9.8	20.25	0.006	363.6	ERR
Allyl Chloride	1	1.00	2.0		10.8	20.25	0.003	312.5	0.6
Allyl Glycidyl Ether (AGE)	10	5.00	10.0		9.1	20.25	0.080	125.0	ERR
Allyl Propryl Disulfide	2 c	2.00	3.0						
alpha-Alumina	b								
Aluminum, total dust	d								
Aluminum	d								
2-Aminoethanol, see Ethanolamine									
2-Aminopyridine	0.5 b	0.50							
Amitrole (solid)									
Ammonia	50		35.0		10.9	20.25	0.112	458.5	0.5
Ammonium Sulfamate (Ammate®)	b								
sec-Amyl Acetate	125 c	125.00							
n-Amyl Acetate	100	100.00			8.0	2.25	0.360	277.8	0.08
Aniline—Skin	5	2.00			9.4	20.25	0.014	577.0	0.2
Anisidine (o,p-Isomers)—Skin	b								
Antimony and Compounds (as Sb)	d								
ANTU (Alpha Naphthylthiourea); 1-(1-naphthyl)-2-thiourea	b								
Arsenic organic compounds (as As)	d								
Arsine	0.05 a	0.05			4.7	20.25		176.4	0.5
Asbestos	b								
Atrazine	b								
Azinphos-methyl—Skin	b								

Compound	OSHA Limit (ppm)	1990 TWA	1990 STEL	Ceiling	Wavelength (Microns)	Pathlength (Meters)	Absorbance (AU)	Linear Term (ppm/AU)	Min DL (ppm)
Barium Sulfate	b								
Barium (Soluble Compounds)	d								
Benomyl	c								
Benzene	10	1.00	5.0	25.0	9.9	20.25	0.006	1616.0	ERR
p-Benzoquinone, see Quinone									
Benzoyl Peroxide (explodes when heated)	b								
Benzyl Chloride	1 a	1.00			9.4	20.25		2052.1	2.5
Beryllium and Beryllium compounds	b	0.002							
Biphenyl, see Diphenyl									
Boron Oxide	b								
Boron Trifluoride	f								
Bromacil	b	1.00							
Bromine	0.1 d	0.10	0.3						
Bromine pentafluoride	0.2 c	0.10							
Bromoform—Skin	0.5	0.50			8.8	20.25	0.002	286.0	0.4
Butadiene (1,3-Butadiene)	1000	in process 6(b) rulemaking			11.1	2.25	0.375	2765.8	ERR
1-Butanethiol, see n-Butyl mercaptan									
2-Butanone		200.00	300.00		8.8	12.75	0.487	453.4	ERR
2-Butoxy Ethanol (Butyl Cellosolve @R)—Skin		25.00			8.9	20.25	0.407	123.9	ERR
tert-Butyl Acetate		200.00			9.9	6.75	0.248	809.0	ERR
sec-Butyl Acetate		200.00			9.9	6.75	0.342	590.5	ERR
n-Butyl Acetate		150.00	200.0		8.2	2.25	0.301	549.0	0.07
Butyl Acrylate (tacky liquid)		10.00			8.5	20.25	0.743	66.0	
sec-Butyl Alcohol	150	100.00			10.1	20.25	0.308	497.8	ERR
tert-Butyl Alcohol		100.00	150.0		8.2	20.25	0.470	212.8	ERR
n-Butyl Alcohol				50.0	9.6	20.25	0.462	227.1	ERR
n-Butyl Glycidyl Ether (BGE)	50	25.00			8.9	20.25	0.550	90.9	ERR
n-Butyl lactate		5.00			8.9	20.25	0.476	208.1	
Butyl Mercaptan	10	0.50			3.4	20.25	0.040	250.0	ERR
p-tert-Butyl Toluene	10	10.00	20.0		12.3	20.25	0.015	666.7	2.0
Butylamine—Skin	5			5.0	13.0	20.25	0.025	200.0	0.8
tert-Butylchromate (as CrO_3)—Skin	b								
o-sec-Butylphenol (Skin)	b	5.00							
Cadmium Dust	d	in process of 6(b) rulemaking							
Cadmium Fume	d	in process of 6(b) rulemaking							
Calcium Carbonate Dust ($CaCO_3$)	b								
Calcium Carbonate Respirable Fraction	b								
Calcium Cyanamide N—CN—Ca	b								
Calcium Hydroxide	b								
Calcium Oxide	b								
Calcium Silicate Respirable Fraction	b								

Calcium Silicate Total Dust	b								
Calcium Sulfate Respirable Fraction	b								
Camphor	b								
Caprolactum dust	b								
Caprolactum vapor	b	5.00	10.0						
Captafol (Difolatin)	b								
Captan	b								
Carbaryl (Sevin @R)	b								
Carbofuran (Furadan)	b								
Carbon Black	b								
Carbon Dioxide	5000	10000.00	30000.0		4.3	0.75	1.327	3524.1	0.4
Carbon Disulfide (skin)	20	4.00	12.0		4.7	20.25	0.018	1181.4	4.8
Carbon Monoxide	50	35.00		200.00	4.7	20.25	0.051	1646.0	2.1
Carbon Tetrabromide	b,g	0.10	0.3						
Carbon Tetrachloride	10	2.00			12.6	20.25	0.031	371.9	0.8
Carbonyl Fluoride		2.00	5.0		10.4	20.25	0.241	406.4	
Catechol—see pyrocatechol (skin)	b		5.0						
Cellosolve Acetate, see 2-Ethoxyethyl Acetate									
Cellulose Respirable Fraction	b								
Cellulose Total Dust	b								
Cesium Hydroxide	b								
Chlordane—Skin	b								
Chlorinated Camphene (Skin)	b								
Chlorinated Diphenyl Oxide	b								
Chlorine	d	0.50	1.0						
Chlorine Dioxide	0.1 c	0.10	0.3						
Chlorine Trifluoride	0.1 c			0.1					
Chloroacetaldehyde	1 e			1.0					
a-Chloroacetophenone (Phenacylchloride)	0.05 b	0.05							
Chloroacetyl Chloride		0.05			12.6	20.25	0.144	1344.0	
Chlorobenzene (Monochlorobenzene)	75	75.00			9.3	20.25	0.229	373.9	ERR
o-Chlorobenzylidene Malononitrile (OCBM)	0.05 c			0.1					
Chlorobromomethane	200	200.00			8.4	12.75	0.364	670.5	ERR
Chlorodifluoromethane (Freon 22)		1000.00			9.2	12.75	0.493	69.7	0.0
Chlorodiphenyl (42% Chlorine)—Skin	b								
Chlorodiphenyl (54% Chlorine)—Skin	b								
2-Chloroethanol, see Ethylene Chlorohydrin									
Chloroform (Trichloromethane)	50	2.00			13.0	6.75	0.606	101.6	ERR
Chloropentafluoroethane (genetron 115)		1000.00			8.1	2.25	0.300	35.7	
Chloropicrin (Trichloronitromethane)	0.1 a	0.10			11.5	20.25		50.0	0.1
Chloroprene (2-Chloro-1,3-butadiene)—Skin	25	10.00			11.4	20.25	0.080	312.5	ERR
Chloropyrifos (Skin)	b								
o-Chlorostyrene		50.00	75.0		13.1	20.25	0.254	773.2	
o-Chlorotoluene		50.00			13.3	20.25	0.438	230.7	
2-Chloro-1,3-Butadiene, see Chloroprene									

Compound	OSHA Limit (ppm)	1990 TWA	1990 STEL	Ceiling	Wavelength (Microns)	Pathlength (Meters)	Absorbance (AU)	Linear Term (ppm/AU)	Min DL (ppm)
1-Chloro-1-nitropropane	20	2.00			12.4	20.25	0.041	494.8	ERR
1-Chloro-2,3-epoxypropane, see Epichlorohydrin									
2-Chloro-6-trichloromethyl pyridine Respirable Fraction	b								
2-Chloro-6-trichloromethyl pyridine Total dust	b								
Chromic Acid and Chromates	b								
Chromium Metal and Insoluble Salts	b								
Chromium Soluble Chromic, Chromous Salts (as Cr)	d								
Clopidol Respirable Fraction	b								
Clopidol Total Dust	b								
Coal Dust < or = 5% SIO_2	b								
Coal Dust > or = 5% SIO_2	b								
Coal Tar Pitch Volatiles (Benzene Soluble Fraction)	b								
Coal Tar Pitch Volatiles (Benzene Solub Anthracene, BaP, P)	b								
Coal Tar Pitch Volatiles (Benzene Solub Pyrene)	b								
Cobalt, Metal Fume and Dust	b								
Cobalt Carbonyl as (CO)	b								
Cobalt Hydrocarbonyl (as CO)	b								
Copper Dusts and Mists	d								
Copper Fume	d								
Cotton Dust (Raw)	b	1.00							
Crag @R Herbicide Total Dust	b								
Crag @R Herbicide Respirable Fraction	b								
Cresol (All Isomers)—Skin	5	5.00			8.9	20.25	0.019	468.4	0.2
Crotonaldehyde	2	2.00			8.7	20.25	0.010	200.0	ERR
Crufomate	b								
Cumene—Skin	50	50.00			9.8	20.25	0.029	1748.2	ERR
Cyanamide	b								
Cyanide (as CN)—Skin	b								
Cyanogen (Ethanedinitrile)—highly poisonous		10.00			4.7	20.25	0.016	6250.0	20.0
Cyanogen chloride	c			0.3					
Cyclohexane	300	300.00			3.4	2.25	0.428	789.7	0.08
Cyclohexanol (Skin)	50	50.00			9.3	20.25	0.274	173.8	ERR
Cyclohexanone (Skin)	50	25.00			8.3	20.25	0.088	559.1	ERR
Cyclohexene	300	300.00			8.8	20.25	0.120	2500.0	4.4

Cyclohexylamine		10.00			9.6	20.25	0.284	1763.1	
Cyclonite	b								
Cyclopentadiene	75 c	75.00							
Cyclopentane		600			11.4	20.25	0.147	4079.8	
Cyhexatin	b								
DDT—Skin	b								
DDVP—Skin	0.1				9.4	20.25	0.002	71.4	0.09
Decaborane—Skin	0.05 e	0.05	0.2						
Demeton—Skin	b								
Diacetone Alcohol (4-Hydroxy-4-methyl-2-pentanone)	50	50.00			8.5	20.25	0.335	149.3	ERR
1,2-Diaminoethane, see Ethylenediamine									
Diazinon (Skin)	c								
Diazomethane	0.2 c	0.20			4.8			380.2	0.6
Diborane	0.1 a	0.10			3.8	20.25			
Dibutyl Phosphate	1 b	1.00	2.0						
2-N-Dibutylaminoethanol		2.00			9.4	20.25	0.058	11972.3	
Dibutylphthalate	b								
Dichloroacetylene	c			0.1					
p-Dichlorobenzene	75	75.00	110.0		9.2	20.25	0.042	2205.6	2.5
o-Dichlorobenzene	50			50.0	13.5	20.25	0.240	255.7	0.6
Dichlorodifluoromethane (F-12)	1000	1000.00			9.3	0.75	0.721	1316.4	0.06
1,1-Dichloroethane	100	100.00			9.5	20.25	0.301	365.3	ERR
Dichloroethyl Ether—Skin	15	5.00	10.0		9.0	20.25	0.208	95.7	0.09
1,2-Dichloroethylene	200	200.00			12.2	2.25	0.297	695.0	ERR
Dichloromethane, see Methylene chloride									
Dichloromonofluoromethane (F-21)	1000	10.00			9.4	0.75	0.460	2125.5	0.08
2,4-D (Dichlorophenoxyacetic acid)	b								
1,2-Dichloropropane, see Propylene dichloride									
1,2-Dichloropropene (Skin)	c	1.00							
2,2-Dichloropropionic Acid	b	1.00							
Dichlorotetrafluoroethane (F-114)	1000	1000.00			8.7	0.75	0.599	1520.1	0.06
1,1-Dichloro-1-nitroethane	10	2.00			9.1	20.25	0.100	100.0	ERR
1,3-Dichloro-5,5-dimethyl Hydantoin	b								
Dicrotophos (Skin)	b								
Dicyclopentadiene		5.00			13.2	20.25	0.198	499.5	
Dicyclopentadienyl Iron Total Dust	b								
Dicyclopentadienyl Iron Respirable Fraction	b								
Dieldrin—Skin	b								
Diethanolamine	b	3.00							
Diethyl Ketone		200.00			9.0	20.25	0.200	250.0	0.4
Diethyl Phthalate	b								
Diethylamine	25	10.00	25.0		9.0	20.25	0.038	542.3	ERR
2-Diethylaminoethanol—Skin	10	10.00			9.4	20.25	0.080	125.0	ERR

Compound	OSHA Limit (ppm)	1990 TWA	1990 STEL	Ceiling	Wavelength (Microns)	Pathlength (Meters)	Absorbance (AU)	Linear Term (ppm/AU)	Min DL (ppm)
Diethylene Triamine	b	1.00							
Diethylether, see Ethyl ether									
Diflurordibromomethane	100	100.00			9.2	2.25	0.360	277.8	0.04
Diglycidyl Ether (DGE)	0.5 c	0.10							
Dihydroxybenzene, see Hydroquinone									
Diisobutyl Ketone	50	25.00			8.6	20.25	0.120	416.7	ERR
Diisopropylamine—Skin	5	5.00			8.5	20.25	0.026	192.3	ERR
Dimethoxymethane, see Methylal									
Dimethyl Acetamide—Skin	10	10.00			10.1	20.25	0.030	478.8	0.4
Dimethylamine	10	10.00			8.8	20.25	0.013	759.1	1.2
Dimethylaminobenzene, see Xylidine									
Dimethylaniline (N,N-Dimethylaniline)—Skin	5	5.00	10.0		8.6	20.25	0.015	333.3	ERR
Dimethylbenzene, see Xylene									
Dimethylformamide—Skin	10	10.00			9.4	20.25	0.072	181.3	ERR
2,6-Dimethylheptanone, see Diisobutyl Ketone									
1,1-Dimethylhydrazine—Skin	0.5	0.50			11.2	20.25	0.004	203.5	ERR
Dimethylphthalate—(Skin)	b								
Dimethylsulfate—Skin	1	0.10			9.9	20.25	0.040	25.0	0.03
Dimethyl-1,2-dibromo-2,2-dichloroethylphosphate (Dibrom®)	b								
Dinitolmide (3,5-Dinitro-o-toluamide)	b								
Dinitrobenzene (All Isomers)—Skin	b								
Dinitrotoluene—Skin	b								
Dinitro-o-cresol—Skin	b								
Dioxane (Diethylene Dioxide)—Skin	100	25.00			9.1	6.75	0.378	253.6	0.1
Dioxathion (Delnav) (Skin)	c								
Diphenyl	0.2 b	0.20							
Diphenylamine	b								
Diphenylmethane Diisocyanate, see Methylene Bisphenyl Isocyanate (MDI)									
Dipropyl ketone		50.00			8.9	20.25	0.259	771.3	
Dipropylene Glycol Methyl Ether—Skin	100	100.00	150.0		9.2	6.75	0.423	256.9	ERR
Diquat	b								
Disulfiram	b								
Disulfoton (Skin)	c								
Diuron	b								
Divinyl benzene		10.00			11.0	20.25	0.098	999.0	
Di-sec-octyl Phthalate (Di-2-ethylhexylphthalate)	b								
2,6-Di-tert-butyl-p-cresol	b								

Emery Respirable Function	b								
Emery Total Dust	b								
Endosulfan—Skin	b								
Endrin—Skin	b								
Epichlorohydrin—Skin	5	2.00			11.8	20.25	0.013	384.6	ERR
EPN—Skin (phenylphosphonothioic acid o-ethylo-p-nitrophen	b								
1,2-Epoxypropane, see Propylene Oxide									
2,3-Epoxy-1-propanol, see Glycidol									
Ethanethiol, see Ethyl Mercaptan									
Ethanolamine	3	3.00	6.0		13.0	20.25	0.004	528.5	ERR
Ethion—Skin	b								
2-Ethoxyethanol—Skin	200	in process 6(b) rulemaking			8.9	6.75	0.613	305.2	ERR
2-Ethoxyethylacetate (Cellosolve® Acetate)—Skin	100	in process 6(b) rulemaking			8.9	6.75	0.366	257.9	0.1
Ethyl Acetate	400	400.00			8.3	0.75	0.292	1300.6	0.07
Ethyl Acrylate—Skin	25	5.0	25.0		8.4	20.25	0.330	75.8	ERR
Ethyl Alcohol (Ethanol)	1000	1000.00			9.7	2.25	0.495	1984.7	ERR
Ethyl Amyl Ketone		25.00			9.0	20.25	0.150	1373.3	
Ethyl Benzene	100	100.00	125.0		9.8	20.25	0.044	2416.6	ERR
Ethyl Bromide	200	200.00	250.0		8.0	6.75	0.219	915.7	ERR
Ethyl Butyl Ketone (3-Heptanone)	50	50.00			9.0	20.25	0.096	533.3	ERR
Ethyl Chloride	1000	1000.00			10.6	6.75	0.393	2635.9	ERR
Ethyl Ether	400	400.00	500.0		8.9	2.25	0.625	708.4	ERR
Ethyl Formate	100	100.00			8.5	2.25	0.300	333.5	0.08
Ethyl Mercaptan	10	0.50			3.3	20.25	0.029	358.1	0.8
Ethyl sec-Amyl Ketone (5-Methyl-3-heptanone)	25				9.0	20.25	0.040	625.0	ERR
Ethyl Silicate	100	10.00			9.1	2.25	0.399	250.6	0.04
Ethylamine	10	10.00			3.4	20.25	0.015	681.5	ERR
Ethylene Chlorohydrin—Skin	5			1.0	9.3	20.25	0.026	192.3	ERR
Ethylene Dibromide	20	in process 6(b) rulemaking			8.6	20.25	0.064	353.2	ERR
Ethylene Dichloride	50	1.00	2.0		8.3	20.25	0.078	682.7	ERR
Ethylene Glycol	b			50.0					
Ethylene Glycol Dinitrate and/or Nitroglycerin	0.2 c,e				12.2	20.25			
Ethylene Glycol Monomethyl Ether Acetate, see Methyl Cellosolve Acetate									
Ethylene Imine—Skin (ethylenimine)	0.5 c								
Ethylene Oxide	1	see 1910.1047			3.3	20.25	0.003	459.3	0.7
Ethylenediamine	10	10.00			13.0	20.25	0.035	285.7	1.2
Ethylidene Norbornene	c			5.0					
Ethylidine Chloride, see 1,1-Dichloroethane									
n-Ethylmorpholine—Skin	20 c	5.00							
Fenamiphos—Skin	b								
Fensulfothion (Dasanit)	b								

Compound	OSHA Limit (ppm)	1990 TWA	1990 STEL	Ceiling	Wavelength (Microns)	Pathlength (Meters)	Absorbance (AU)	Linear Term (ppm/AU)	Min DL (ppm)
Fenthion—Skin	b								
Ferbam Respirable Fraction	b								
Ferbam Total Dust	b								
Ferrovanadium Dust	b								
Fluoride (as Dust)	b	2.50							
Fluoride (as F)	b	2.50							
Fluorine	0.1 d	0.10							
Fluorocarbon 11, see Fluorotrichloromethane									
Fluorocarbon 112, see 1,1,2,2-Tetrachloro-1,2-difluoroethane									
Fluorocarbon 112A, see 1,1,1,2-Tetrachloro-2,2-Difluoroethane									
Fluorocarbon 113, see 1,1,2-Trichloro-1,2,2-trifluoroethane									
Fluorocarbon 114, see Dichlorotetrafluoroethane									
Fluorocarbon 12, see Dichlorodifluoromethane									
Fluorocarbon 13B1, see Trifluoromonobromomethane									
Fluorocarbon 21, see Dichloromonofluoromethane									
Fluorotrichloromethane (F-11)	1000			1000.0	11.0	6.75	0.450	2317.3	1.4
Fonofos—Skin	b								
Formaldehyde	1	see 1910.1048	2.0		3.6	20.25	0.012	316.0	0.5
Formamide	b	20.00	30.0						
Formic Acid	5	5.00			9.4	20.25	0.032	214.8	ERR
Furfural—Skin	5	2.00			13.3	20.25	0.050	100.0	ERR
Furfuryl Alcohol	50	10.00	15.0		9.8	20.25	0.090	555.6	ERR
Gasoline	b	300.00	500.00						
Germanium tetrahydride	c	0.20							
Glutaraldehyde (Pentanediol)	1			0.2	3.7	20.25	0.007	1484.0	3.2
Glycerin (mist) Respirable Fraction	b								
Glycerin (mist) Total Dust	b								
Glycidol (2,3-Epoxy-1-propanol)	50	25.00			9.9	20.25	0.019	2676.4	ERR
Glycol Monoethyl Ether, see 2-Ethoxyethanol									
Grain Dust (oat, wheat, barley)	b								
Graphite (natural) respirable dust	b								
Graphite (synthetic) respirable fraction	b								
Graphite (synthetic) total dust	b								
Guthion®, see Azinphosmethyl									

Gypsum respirable fraction	b								
Gypsum total dust	b								
Hafnium	b								
Heptachlor—Skin	b								
Heptane (n-Heptane)	500	400.00	500.0		3.4	0.75	0.285	1841.7	ERR
Hexachlorobutadiene		0.02			11.6	20.25	0.408	240.5	
Hexachlorocyclopentadiene		0.01			12.2	20.25	0.190	1053.2	
Hexachloroethane—Skin	1	1.00			12.8	20.25	0.040	25.0	0.09
Hexachloronaphthalene—Skin	b								
Hexafluoroacetone—Skin		0.10			10.3	20.25	0.479	125.2	
Hexane Isomers	c	500.0	1000.0						
Hexane (n-Hexane)	500	50.00			3.4	0.75	0.254	2138.4	0.1
2-Hexanone	100	5.00			8.6	20.25	0.420	238.1	ERR
Hexone (Methyl Isobutyl Ketone)	100	50.00	75.00		8.5	20.25	0.327	301.5	ERR
sec-Hexyl Acetate	50	50.00			9.6	20.25	0.204	259.0	ERR
Hexylene glycol				25.0	7.1	20.25	0.138	3560.6	
Hydrazine—Skin	1	0.10			10.5	20.25	0.002	733.2	0.6
Hydrogen Bromide	3 f			3.0					
Hydrogen Chloride	5			5.0	3.4	20.25	0.002	2268.8	4.8
Hydrogen Cyanide—Skin	10		4.7		3.0	20.25	0.020	583.9	1.7
Hydrogen Fluoride	3 a	3.00	6.0		2.5	20.25		1941.7	10.0
Hydrogen Peroxide (90%)	1 a	1.00			7.8	20.25		769.2	1.3
Hydrogen Selenide	0.05 g	0.05							
Hydrogen Sulfide	20 a	10.00	15.0		8.2	20.25		118000.0	270
Hydrogenated terphenyls	c	0.50							
Hydroquinone	b								
2-Hydroxypropyl Acrylate—Skin	c	0.50							
Indene (Indonaphthene)		10.00			13.0	20.25	0.230	881.5	
Indium and compounds (as In)	b								
Iodine	0.1 d	0.10							
Iodoform—triiodomethane	b	0.60							
Iron Oxide Fume	b								
Iron Pentacarbonyl (as Fe)	d	0.1		0.2					
Iron salts (soluble)	b								
Isoamyl Acetate	100	100.00			9.4	12.75	0.353	283.2	ERR
Isoamyl Alcohol	100	100.00	125.0		9.4	12.75	0.288	347.6	ERR
Isobutyl Acetate	150				8.2	2.25	0.400	375.0	ERR
Isobutyl Alcohol	100	50.00			9.6	12.75	0.464	215.7	ERR
Isooctyl Alcohol—Skin vapors are poisonous!	c	50.00							
Isophorone	25	4.00			3.4	20.25	0.120	208.3	0.5
Isophorone Diisocyanate	b	0.001	0.02						
2-Isopropoxyethanol		25.00			9.5	20.25	0.738	139.7	
Isopropyl Acetate	250	250.00	310.0		8.0	2.25	0.629	342.0	ERR
Isopropyl Alcohol	400	400.00	500.0		8.9	6.75	0.378	1146.4	ERR

Compound	OSHA Limit (ppm)	1990 TWA	1990 STEL	Ceiling	Wavelength (Microns)	Pathlength (Meters)	Absorbance (AU)	Linear Term (ppm/AU)	Min DL (ppm)
Isopropyl Ether	500	500.00			9.0	2.25	0.549	981.0	ERR
Isopropyl Glycidyl Ether (IGE)	50 c	50.00	75.0						
Isopropylamine	5	5.00	10.0		9.0	20.25	0.009	553.0	ERR
N-Isopropylaniline	c	2.00							
Kaolin Respirable Fraction	b								
Koalin Total Dust	b								
Ketene	0.5 c	0.50	1.5						
Lead inorganic	see 1910.1025								
Limestone Respirable fraction	b								
Limestone Total Dust	b								
Lindane—Skin (hexachlorocyclohexane)	b								
Lithium Hydride	b								
LPG (Liquified Petroleum Gas)	1000	1000.00			3.4	0.75	0.240	4166.7	ERR
Magnesite respirable fraction	b								
Magnesite total dust	b								
Magnesium Oxide Fume Respirable fraction	b								
Magnesium Oxide Fume Total Dust	b								
Malathion—Skin Total Dust	b								
Malathion—Skin Respirable Fraction	b								
Maleic Anhydride	0.25 b	0.25							
Manganese	b								
Manganese cyclopentadienyl tricarbonyl—Skin	b								
Maganese Tetraoxide	b								
Marble Respirable fraction	b								
Marble Total Dust	b								
Mercury	d								
Mesityl Oxide	25	15.00	25.0		8.2	20.25	0.060	416.7	ERR
Methacrylic acid—Skin Ha		20.00			8.9	20.25	0.556	354.4	
Methanethiol, see Methyl mercaptan									
Methomyl (Lannate)	b								
Methoxychlor Respirable Fraction	b								
Methoxychlor Total Dust	b								
2-Methoxyethanol, see Methyl Cellosolve									
4-Methoxyphenol	b								
Methyl 2-Cyanoacrylate	c	2.00	4.0						
Methyl Acetate	200	200.00	250.0		9.6	6.75	0.330	624.4	0.3
Methyl Acetylene (Propyne)	1000	1000.00			3.0	6.75	0.402	3121.2	2.8
Methyl Acetylene-Propadiene Mixture (MAPP)	1000 c	1000.00	1250.0						
Methyl Acrylate—Skin	10	10.00			8.5	20.25	0.162	75.5	ERR

Methyl Alcohol (Methanol)	200	200.00	250.0		9.7	6.75	0.302	747.8	ERR
Methyl Amyl Alcohol, see Methyl Isobutyl carbinol									
Methyl Bromide—Skin	20	5.00			7.5	20.25	0.018	1100.0	ERR
Methyl Butyl Ketone, see 2-Hexanone									
Methyl Cellosolve® Acetate—Skin (2-Methoxyethyl acetate)	25	In process of 6(b) rulemaking			8.0	20.25	0.392	61.7	ERR
Methyl Cellosolve®—Skin (2-Methoxyethanol)	25	In process of 6(b) rulemaking			9.6	20.25	0.120	233.2	0.3
Methyl Chloride	100	50.00	100.0		13.5	20.25	0.162	625.4	3.0
Methyl Chloroform (1,1,1-Trichloroethane)	350	350.00	450.0		9.3	2.25	0.313	1179.0	ERR
Methyl Demeton—Skin	b								
Methyl Ethyl Ketone Peroxide (MEKP)	b,c			0.7					
Methyl Ethyl Ketone (MEK), see 2-Butanone									
Methyl Formate	100	100.00	150.0		8.5	2.25	0.396	253.0	0.06
Methyl Hydrazine(Mono-methyl hydrazine)—Skin	0.2 a				10.9	20.25	0.001	321.0	0.6
Methyl Iodide—Skin	5	2.00			3.4	20.25	0.005	1283.1	1.8
Methyl Isoamyl Ketone		50.00			8.6	20.25	0.600	83.3	0.2
Methyl Isobutyl Carbinol—Skin	25	25.00	40.0		8.7	20.25	0.053	471.7	ERR
Methyl Isobutyl Ketone, see Hexone									
Methyl Isocyanate—Skin	0.02 a	0.02			4.4	20.25		19.2	0.1
Methyl Isopropyl Ketone		200.00			8.8	20.25	0.100	500.0	0.8
Methyl Mercaptan	10	0.50			3.4	20.25	0.016	796.2	1.5
Methyl Methacrylate	100	100.00			8.7	6.75	0.576	187.5	ERR
Methyl n-Amyl Ketone	100	100.00			8.6	20.25	0.253	394.7	ERR
Methyl Parathion—Skin	b								
Methyl Propyl Ketone, see 2-Pentanone									
Methyl Silicate	b	1.00							
alpha-Methylstyrene	100	50.00	100.0		11.3	20.25	0.302	319.5	ERR
Methylacrylonitrile—Skin	c	1.00							
Methylal (Dimethoxymethane)		1000.00			9.5	0.75	0.437	241.8	0.1
Methylamine—see methyl isobutyl carbinol									
Methylcyclohexane	500		400.0		3.4	0.75	0.281	1670.2	ERR
Methylcyclohexanol	100		50.0		9.5	20.25	0.240	416.7	ERR
o-Methylcyclohexanone—Skin	100	50.00	75.0		8.9	20.25	0.180	555.6	ERR
Methylcyclopentadienyl manganese tricarbonyl (as Mn)	b								
4,4-Methylene bis (2-chloroaniline) (MBOCA)—Skin	b,c	0.02							
Methylene bis (4-cyclohexylisocyanate)	c			0.01					
Methylene Bisphenyl Isocyanate (MDI)	0.02 b			0.02					
Methylene Chloride	500	In process of 6(b) rulemaking			13.5	0.75	0.245	2186.1	ERR
Metribuzin	b								

Compound	OSHA Limit (ppm)	1990 TWA	1990 STEL	Ceiling	Wavelength (Microns)	Pathlength (Meters)	Absorbance (AU)	Linear Term (ppm/AU)	Min DL (ppm)
Molybdenum Insoluble Compounds	b								
Molybdenum Soluble Compounds	b								
Monocrotophos (Azodrin)	b								
Monomethyl Hydrazine—Skin	0.2 a			0.2	10.9	20.25		321.0	0.6
Monomethylaniline—Skin	2	0.50			9.4	20.25	0.002	1431.9	1.9
Morpholine—Skin	20	20.00	30.0		9.1	20.25	0.048	366.2	ERR
Naphtha (Coaltar)	100 b	100.00							
Naphthalene	10 b	10.00	15.0						
Nickel Carbonyl	0.001 a	0.001			4.9	20.25		34.0	0.2
Nickel Metal and Insoluble Compounds as Ni	b								
Nickel Metal and Soluble Compounds as Ni	b								
Nicotine—Skin	b								
Nitric Acid	2 f	2.00	4.0		7.6				
Nitric Oxide—(NO)	25 h	25.00			5.3	20.25	0.032	1108.1	3.9
p-Nitroaniline—Skin	1 b								
Nitrobenzene—Skin	1	1.00			11.9	20.25	0.002	820.6	0.9
p-Nitrochlorobenzene—Skin	b								
4-Nitrodiphenyl	c	see 1910.1003							
Nitroethane	100	100.00			9.0	20.25	0.076	1336.7	ERR
Nitrogen Dioxide (NO_2)	5		1.0		6.3	20.25	0.065	76.9	0.4
Nitrogen Trifluoride	10	10.00			11.0	20.25	0.350	28.6	0.05
Nitroglycerin—Skin	0.2 c,e								
Nitromethane	100	100.00			3.4	20.25	0.048	2079.5	ERR
1-Nitropropane	25	25.00			12.5	20.25	0.017	1470.5	4.8
2-Nitropropane	25	10.00			11.8	20.25	0.027	933.8	2.3
N-Nitrosodimethylamine		see 1910.1016			9.9	20.25	0.442	106.3	
p-Nitrotoluene—Skin					11.8	20.25	0.007	1085.4	1.8
o-Nitrotoluene—Skin					13.8	20.25	0.102	994.1	
m-Nitrotoluene—Skin	5	2.0			7.3	20.25	0.144	1393.7	
Nitrotrichloromethane, see Chloropicrin									
Nonane		200.00			3.4	20.25	0.683	74.7	
Octachloronaphthalene—Skin	b								
Octane	500	300.00	375.00		3.4	0.75	0.316	1732.6	ERR
Oil Mist, Mineral	b								
Organo(alkyl) Mercury	b								
Osmium Tetroxide	b	0.0002	0.0006						
Oxalic Acid	b								
Oxygen Difluoride	0.05 f			0.05					
Ozone	0.1 c	0.10	0.3						
Paraquat—Skin	b								

Parathion—Skin	b								
Pentaborane	0.005 e	0.0050	0.0150						
Pentachloronaphthalene—Skin	b								
Pentachlorophenol—Skin	b								
Pentaerythritol Respirable Fraction	b								
Pentaerythritol Total Dust	b								
Pentane	1000	600.00	750.0		3.4	0.75	0.423	2451.1	0.2
2-Pentanone	200	200.00	250.0		8.5	6.75	0.298	685.5	ERR
Perchloroethylene		25.00			12.7	20.25	0.370	275.8	
Perchloromethyl Mercaptan	0.1 a	0.10			13.2	20.25		90.3	0.3
Perchloryl Fluoride	3 e	3.00	6.0						
Perlite Respirable Fraction	b								
Perlite Total Dust	b								
Petroleum Distillates (Naphtha)	500	400.00			3.4	2.25	0.650	769.2	ERR
Phenol—Skin	5	5.00			8.5	20.25	0.006	1055.2	1.7
Phenothiazine—Skin	b								
Phenyl Ether (Vapor)	1	1.00			8.0	20.25	0.013	76.9	ERR
Phenyl Ether-Biphenyl Mixture (Vapor)	1	1.00			8.1	20.25	0.011	90.9	ERR
Phenyl Glycidyl Ether (PGE)	10 c	1.00							
Phenyl mercaptan		0.50			13.6	20.25	0.301	338.9	
p-Phenylene Diamine—Skin	b								
Phenylethylene, see Styrene									
Phenylhydrazine—Skin	5	5.00	10.0		8.5	20.25	0.003	2546.3	ERR
Phenylphosphine				0.05					
Phorate—Skin	b								
Phosdrin (Mevinphos)—Skin	b	0.01	0.03						
Phosgene (Carbonyl Chloride)	0.1	0.10			11.9	20.25	0.002	27.5	0.1
Phosphine	0.3 a	0.30			10.1	20.25		1000.0	1.4
Phosphoric Acid	b								
Phosphorus Oxychloride—Fuming Liquid		0.10			7.5	20.25	0.577	194.1	
Phosphorus Pentachloride	b								
Phosphorus Pentasulfide	b								
Phosphorus Trichloride	0.5 g	0.20	0.5						
Phosphorus (Yellow)	b								
Phthalic Anhydride	2 b	1.00							
m-Phthalodinitrile	b								
Picloram Respirable Fraction	b								
Picloram Total Dust	b								
Picric Acid—Skin	b								
Pidone® (2-Pivalyl-1,3-Indandione)	b								
Piperazine Dihydrochloride	b								
Plaster of Paris Respirable Fraction	b								
Plaster of Paris Total Dust	b								
Platinum (Soluble Salts) as Pt	b								

Compound	OSHA Limit (ppm)	1990 TWA	1990 STEL	Ceiling	Wavelength (Microns)	Pathlength (Meters)	Absorbance (AU)	Linear Term (ppm/AU)	Min DL (ppm)
Portland Cement Respirable fraction	b								
Portland Cement Total Dust	b								
Propane	1000	1000.00			3.4	0.75	0.304	3568.8	ERR
Propargyl Alcohol → 2-Propyn-1-ol		1.00			9.7	20.25	0.533	180.2	
beta-Propiolactone → 2-oxetanone		see 1910.1013			9.2	20.25	0.487	205.4	
Propionic Acid		10.00			8.8	20.25	0.628	163.9	
Propoxur (Baygon)	b								
n-Propyl Acetate	200	200.00	250.0		8.1	2.25	0.540	370.4	ERR
Propyl Alcohol	200	200.00	250.0		9.6	6.75	0.288	756.2	ERR
n-Propyl Nitrate	25	25.00	40.0		10.4	20.25	0.107	234.5	ERR
Propylene Dichloride	75	75.00	110.0		9.9	20.25	0.123	639.0	ERR
Propylene Glycol Dinitrate	c	0.05							
Propylene Glycol Monomethyl ether		100.00	150.0		9.0	20.25	0.446	112.2	
Propylene Imine—Skin	2 c	100.00	150.0						
Propylene Oxide	100	20.00			12.1	20.25	0.245	421.5	ERR
Propyne, see Methyl Acetylene									
Pyrethrum (Insect powder; Insecticide prep's)	b								
Pyridine	5	5.00			14.2	20.25	0.050	100.0	ERR
Quinone	0.1 b	0.10							
Resorcinol—1,3-benzenediol	b	10.00	20.0						
Rhodium, Metal Fume and Dusts, as Rh	d								
Rhodium, Metal Fume and Dusts	d								
Rhodium, Metal Fume and Dusts, as Soluble Salts	d								
Ronnel	b								
Rotenone (Commercial)	b								
Rouge Respirable Fraction	b								
Rouge Total Dust	b								
Selenium Compounds (as Se)	d								
Selenium Hexafluoride	0.05 a	0.05			12.9	20.25	0.003	19.2	0.07
Silica (various forms)	b	See Table Z-3							
Silicon	b								
Silicon Carbide Respirable Fraction	b								
Silicon Carbide Total Dust	b								
Silicon Tetrahydride	b	5.00							
Silver Metal and Soluble Compounds	d								
Sodium Azide (as HN_3)—Skin	b			0.1					
Sodium Azide (as NaN_3)—Skin	b								
Sodium Fluoroacetate (1080)—Skin	b								
Sodium Hydroxide	b								
Sodium metabisulfite—sodium pyrosulfite	b								

Starch Respirable Fraction	b								
Starch Total Dust	b								
Stibine	0.1 c	0.10							
Stoddard Solvent	500	100.00			3.4	0.75	0.339	1448.9	ERR
Strychnine	b								
Styrene	100	50.00	100.0		11.2	20.25	0.295	456.8	0.5
Subtilisins (Proteolytic Enzymes)	b								
Sucrose Respirable Fraction	b								
Sucrose Total Dust	b								
Sulfur Dioxide	5		2.0	5.0	8.8	20.25	0.004	1572.8	ERR
Sulfur Hexafluoride	1000	1000.00			10.7	0.75	1.187	723.1	0.01
Sulfur Monochloride	1 g			1.0					
Sulfur Pentafluoride	0.025 c			0.01					
Sulfuric Acid	b								
Sulfuryl Fluoride	5	5.00	10.0		11.5	20.25	0.100	50.0	ERR
Sulprofos	b								
Systox, see Demeton									
Tantalum	b								
TEDP—Skin	b								
Tellurium	b								
Tellurium Hexafluoride	0.02 a	0.02			13.4	20.25		18.1	0.08
Temephos Respirable Fraction	b								
Temephos Total Dust	b								
TEPP—Skin	b								
Terphenyls	1 b			0.5					
1,1,2,2-Tetrachloroethane	5	1.00			8.5	20.25	0.007	724.2	1.4
Tetrachloroethylene, see Perchloroethylene									
Tetrachloromethane, see Carbon Tetrachloride									
Tetrachloronaphthalene—Skin	b								
1,1,2,2-Tetrachloro-1,2,-Difluoroethane (F-112)	500	500.00			9.8	6.75	0.459	1089.0	ERR
1,1,1,2-Tetrachloro-2,2-Difluoroethane (F-112A)	500 c	500.00							
Tetraethyl Lead (as Pb)—Skin	b								
Tetrahydrofuran	200	200.00	250.0		9.4	6.75	0.379	577.7	ERR
Tetramethyl Lead (as Pb)—Skin	b								
Tetramethyl Succinonitrile—Skin	0.5 c	0.50							
Tetranitromethane	1 e	1.00							
Tetrasodium Pyrophosphate	b								
Tetryl (2,4,6-Trinitrophenylmethylnitramine)—Skin	b								
Thallium (Soluble Compounds)—Skin, as Th	d								
4,4′-Thiobis(6-tert,Butyl-m-cresol) Respirable Fraction	b								

Compound	OSHA Limit (ppm)	1990 TWA	1990 STEL	Ceiling	Wavelength (Microns)	Pathlength (Meters)	Absorbance (AU)	Linear Term (ppm/AU)	Min DL (ppm)
4,4′-Thiobis(6-tert,Butyl-m-cresol) Total Dust	b								
Thioglycolic Acid—Skin (Mercaptoacetic acid)	b	1.00							
Thionyl Chloride	c			1.0					
Thiram	b								
Tin Oxide (as Sn)	b								
Tin (Inorganic Cmpds, Except Oxides)	d								
Tin (Organic Compounds)	d								
Titanium Dioxide Respirable Fraction	b								
Titanium Dioxide Total Dust	b								
Toluene	200	100.00	150.0		13.9	6.75	0.357	682.6	ERR
Toluene-2,4-diisocyanate	0.02 g	0.005	0.0						
p-Toluidine—Skin		2.00			12.3	20.25	0.819	434.8	
m-Toluidine—Skin		2.0			7.6	20.25	0.099	9878.4	
o-Toluidine—Skin	5	5.00			13.5	20.25	0.028	177.4	ERR
Toxaphene, see Chlorinated Camphene									
Tremolite, see Silicates	b								
Tributyl Phosphate	b	0.20							
Trichloro Acetic Acid	b	1.00							
1,1,1-Trichloroethane, see Methyl Chloroform									
1,1,2-Trichloroethane—Skin	10	10.00			10.8	20.25	0.021	484.9	ERR
Trichloroethylene	100	50.00	200.0		10.7	6.75	0.284	434.1	ERR
Trichloromethane, see Chloroform									
Trichlornaphthalene—Skin	b								
Trichloronitromethane, see Chloropicrin									
1,2,3-Trichloropropane	50	10.00			12.4	20.25	0.160	312.5	ERR
1,1,2-Trichloro-1,2,2,-trifluoroethane (F-113)	1000	1000.00	1250.0		8.7	0.75	0.384	2858.8	0.2
Triethylamine	25	10.00	15.0		9.3	20.25	0.078	316.6	ERR
Trifluoromonobromomethane (F-1381)	1000	1000.00			8.5	0.75	0.497	1684.3	0.01
Trimellitic Anhydride	b	0.005							
1,2,4-Trimethyl Benzene		25.00			12.4	20.25	0.143	687.2	
Trimethyl Phosphite		2.00			9.7	20.25	0.715	41.4	
Trimethylamine		10.00	15.0		12.3	20.25	0.243	1644.1	
2,4,6-Trinitrophenyl, see Picric Acid									
2,4,6-Trinitrophenylmethylnitramine, see Tetryl									
Trinitrotoluene—Skin	b								
Triorthocresyl Phosphate	b								
Triphenyl Amine	b								
Triphenyl Phosphate	b								
Tungsten Insoluble Compounds	b								
Tungsten Soluble Compounds	b								
Turpentine	100	100.00			3.4	6.75	0.390	256.4	ERR

Uranium (Insoluble Compounds)	d							
Uranium (Soluble Compounds)	d							
n-Valeraldehyde (Pentanal)		50.00		3.4	20.25	0.578	169.5	
Vanadium V_2O_5 Fume	d							
Vegetable Oil Mist Respirable Fraction	b							
Vegetable Oil Mist Total Dust	b							
Vinyl Acetate		10.00	20.0	8.4	20.25	0.273	35.5	0.1
Vinyl Benzene, see Styrene								
Vinyl Bromide		5.00		10.9	20.25	0.096	520.8	0.9
Vinyl Chloride	1	see 1910.1017		11.3	20.25	0.003	489.1	0.8
Vinyl Cyanide, see Acrylonitrile								
Vinyl Cyclohexene Dioxide	c	10.00						
Vinyl Toluene	100	100.00		11.1	20.25	0.200	500.0	ERR
Vinylidene Chloride (1,1-Dichloroethylene)		1.00		9.2	20.25	0.040	41.7	0.3
VM & P Naptha	c	300.00	400.0					
Warfarin	b							
Welding Fumes (total particulate)	b							
Wood Dust (all soft and hard, except Western red cedar)	b							
Wood Dust (Western red cedar)	b							
m-Xylene alpha, alpha-diamine—Skin	b							
Xylenes (o-, m-, p-isomers)	100	100.00	150.0	13.1	20.25	0.225	478.5	1.9
Xylidine—Skin	5	2.00		7.2	20.25	0.056	89.1	ERR
Yttrium	b							
Zinc Chloride Fume	b							
Zinc Chromate (as CrO_3)	b							
Zinc Oxide Respirable Fraction	b							
Zinc Oxide Total Dust	b							
Zinc Oxide Fume	b							
Zinc Stearate Respirable Fraction	b							
Zinc Stearate Total Dust	b							
Zirconium Compounds (as Zr)	b							

OSHA CHART REVISION 2—OSHA Limits from OSHA Safety and Health Standards (29CFR 1910, OSHA 2206)

Notes: TWA = Time Weighted Average
STEL = Short-term Exposure Limit
Min DL = Minimum Detectable Level (based on path length of 20.25 meters)
a = Absorbance below noise level at OSHA-TWA value
b = Exists as a particulate for vapor pressure concentration
c = Difficult or impossible to obtain commercially
d = Infrared inactive
e = Extremely hazardous
f = Potential analyzer damage
g = Significant carbon dioxide interference
h = Compound analyzed in nitrogen

Appendix E: Answers to All Exercises

APPENDIX E
Answers to All Exercises

Unit 2

Exercise 2-1.

Parts per million, or grains per standard cubic foot.

Exercise 2-2.

A "major source" is defined by the EPA as 28 specific types of facilities, such as kraft pulp and steel mills and other sources, that emit 250 tons per year or more of regulated pollutants.

Exercise 2-3.

Sulfur dioxide emissions can be reduced by using a limestone injection system, by burning low sulfur fuels, or by using wet scrubber systems.

Exercise 2-4.

Low leakage valve designs should be used, such as bellows seals, live loading packings, or diaphragm valves. Also, ensure that adequate gasketing has been provided at all flanged pipe fittings.

Exercise 2-5.

See Figure 2-1 for packing components. The packing box is the cavity into which the packing material is compressed. The packing material itself shapes itself to the surrounding surfaces to seal off the area around the valve stem and prevent leaks. The packing nut or retainer compresses the packing material to press tightly against all surfaces and create a gas- tight seal. It maintains constant pressure on the packing.

Bellows seals create a flexible metallic containment for any gases that escape around the valve stem packing.

Diaphragm valves, by design, fully isolate the valve stem from the process.

All of the above methods are very effective in nearly eliminating fugitive emissions. However, none provide an absolute guarantee that leaks will not occur.

Exercise 2-6.

Male/female joint, tongue and groove, and ring joint.

Exercise 2-7.

One of the most effective analyzers that can be used to monitor waste-burning emissions is infrared absorption, due to its successful use on a wide range of elements and compounds. In addition, acid gas measurement, opacity monitors, and/or particle analyzers would be used.

Exercise 2-8.

See sketch in Figure 2-17. In situ detection measures the gas directly in the stream under actual conditions. Extraction units remove gas from the stream. Therefore, the conditions at the point of measurement are not the same. However, extraction units can be located in much more accessible locations for maintenance service. They are not subjected to the harsh conditions that the in situ unit must survive in.

Exercise 2-9.

Dispersion analysis predicts the amount of dispersion of gas emissions that will occur from a given source. This allows ground level concentrations of pollutants to be predicted.

Air quality analyzers, capable of detecting and measuring levels of pollutants and particulate levels.

Exercise 2-10.

SOx control—Flue gas desulfurization using limestone injection, wet or dry scrubbers with a limestone slurry.

NOx control—Low-NOx burners, flue gas recirculation, ammonia injection, selective catalytic reduction.

Particulate—Electrostatic filters, fabric filters, wet scrubbers, cyclones, dust collectors.

Exercise 2-11.

Refer to sketch in Figure 2-22.

Exercise 2-12.

This highly efficient system for removal of NOx uses ammonia injection to create a chemical reaction between ammonia and the sulfur that results in the creation of nitrogen and water, which are harmless to the environment. Platinum in the catalytic convertor acts as a catalyst, allowing the reaction to occur at a lower gas temperature.

The potential exists, if not tightly controlled, for ammonia to be emitted to atmosphere. The reaction with sulfur dioxide occurs within a narrow temperature range. Gas temperatures emitted from a plant can change rapidly with sudden changes in load or fuel conditions.

Exercise 2-13.

Precombustion control, such as flue gas recirculation for NOx control, occurs before or during the combustion process.

Postcombustion control occurs after combustion, such as flue gas desulfurization (FGD) to limit emission levels.

Unit 3

Exercise 3-1.

Inadequate fresh air entering the building, outgassing by materials in the building, and very high humidity conditions.

Exercise 3-2.

Particle counters are very effective. They are used for air quality analysis in critical areas such as clean rooms.

Exercise 3-3.

Refer to the sketch in Figure 3-13.

Exercise 3-4.

Combustible gas monitors are used to measure levels of flammable gases in relation to their LEL (lower explosive limit). They are used extensively in plants where the potential exists for gas concentrations to collect to a level that could create a hazardous situation. As the concentration nears the LEL, an alarm is activated to warn personnel in the area of a potential hazard. These monitors are typically equipped with sensors for a specific gas type.

Exercise 3-5.

The advantage is in their low cost and ability to supplement current methods of ventilation control. The disadvantage is that they measure only one gas, such as CO_2 or oxygen. This doesn't tell the whole story of air quality. Therefore, they do not effectively ensure good air quality control.

Exercise 3-6.

Fabric filters and wet scrubber systems both can effectively filter particulate from the air. The advantage of fabric filters is their low maintenance requirements as compared to wet scrubbers. Wastewater and sludge must be disposed of regularly with wet systems. However, wet systems must be used when dealing with flammable or explosive substances, such as magnesium dust.

Exercise 3-7.

Refer to the sketch in Figure 3-21.

Exercise 3-8.

Dehumidification increases the energy costs associated with the HVAC system. However, it allows relatively tight control of humidity conditions, reducing the potential for bacterial growth and corrosion within the building.

Exercise 3-9.

Refer to sketch in Figure 3-5.

Exercise 3-10.

A desiccant dryer downstream of a mechanical (refrigerant) dehumidification system will effectively provide very low humidity conditions when needed for a special environment such as a laboratory or humidity-sensitive manufacturing process or storage area.

Unit 4

Exercise 4-1.

pH is the unit of measure that represents the acidity or alkalinity of a substance. It is representative of the ratio of positive hydroxide ions to negative hydroxide ions.

Exercise 4-2.

Batch treatment ensures adequate mixing of the reagent with the wastewater. It also makes sampling of the treated wastewater easier; therefore, less instrumentation is needed for measurement and control.

Continuous flow pH treatment can be used successfully on waste streams where very small variations in pH and temperature occur.

Exercise 4-3.

The point at which sampling occurs *must* be representative of the average stream conditions; it must provide good repeatability and reliability. The point at which effluent discharges into the environment or sanitary waste system is the most critical location in most cases. It is most representative of the water's condition at the point of discharge.

Exercise 4-4.

Volumetric flow rate, pH, turbidity, chlorine concentration, temperature, conductivity (suspended solids).

Exercise 4-5.

Turbidity is the ability of light to pass through a representative sample of water. Suspended particles in the water disperse the beam of light as it passes through the sample.

Exercise 4-6.

Grab samples are cooled to a low temperature to cause stasis to occur. Stasis suspends biological activity in the sample in order to preserve it in its present state until analysis.

Exercise 4-7.

Open channel flow is liquid flow through a partially filled channel, conduit, or trench.

A weir is used in open channel flow applications to create an upstream head condition at a location where the cross-sectional dimensions of the flow area are known (such as the height of the "V" in a V-notch weir). Variations in the height (or head) are representative of the flow rate.

By creating a flow restriction at a known fixed cross-sectional area of a weir or flume and measuring the height of the stream of water passing through that area, the representative flow rate can be calculated.

Exercise 4-8.

Ratio flow control is the measurement of a process flow and proportional (or ratio) flow regulation of a second process flow rate in response to the first. An example would be control of the discharge rate of an effluent into a river based on the river's rate of flow (which will invariably vary with seasonal conditions).

Exercise 4-9.

The amount of reagent added is determined by the deviation of the untreated wastewater from the desired pH value. Reagent addition varies by a factor of ten to the variation in pH from set point.

Exercise 4-10.

High pH indicates an alkaline (base) condition. An acid, such as sulfuric or hydrochloric acid, is added to neutralize the water.

Exercise 4-11.

pH and its measurement are affected by temperature variations. The addition of reagent, as mentioned in exercise 4-9, is at a factor of ten. Therefore, if measurement is off by a pH factor of one, ten times (or one-tenth) the amount of reagent really needed for neutralization is added.

Exercise 4-12.

Activated sludge is a sludge-like material made up of microbiological organisms that digest waste materials, converting the waste into material more easily separated from the waste stream. Oxygen, by aeration (creation of numerous small air bubbles in the water), is needed by the microorganisms in order for them to survive and thrive on waste material.

Exercise 4-13.

Chemical recovery reduces the concentrations of chemicals discharged into the environment. The reclaimed chemicals or compounds can be reused in processes or resold to defray the cost of operating the recovery system.

Exercise 4-14.

Lagoons are a relatively inexpensive way to temporarily store effluent. In addition, if the lagoon has a permeable liner, water will leach out of the lagoon into the surrounding soil. The remaining solids can be removed by a backhoe or shovels and disposed of more easily than wastewater.

Unit 5

Exercise 5-1.

The basic elements of the hydrological cycle are evaporation, precipitation, and migration of the water through rivers, lakes, oceans, the soil and aquifers.

Exercise 5-2.

See Figure 5-3.

Exercise 5-3.

The surrounding topography and level of the water table or aquifer. In addition, wells must be located in an accessible area where they do not pose a potential safety hazard.

Exercise 5-4.

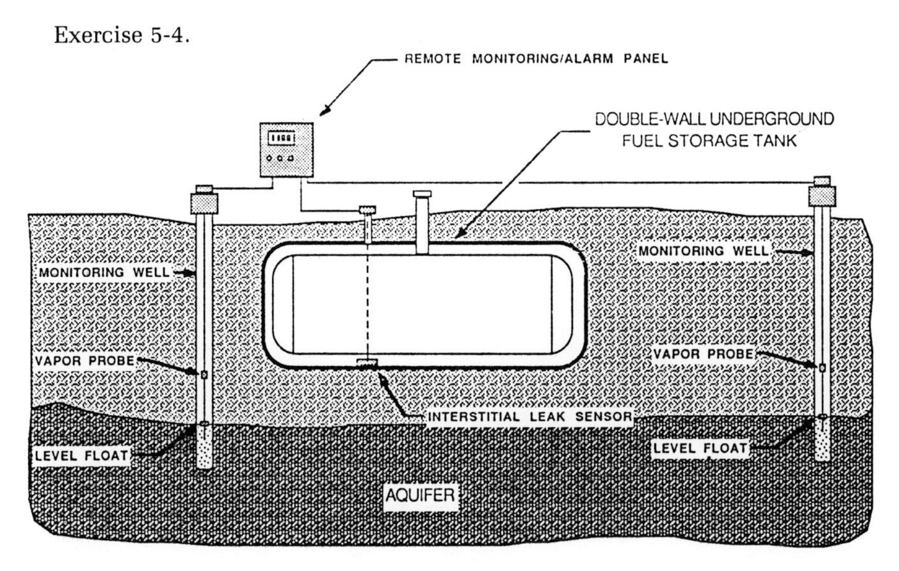

Exercise 5-5.

In a wet area, it is very important that the leak detection sensors be capable of distinguishing between water and the leaking material. A well with a float and vapor detection sensor works well in wet areas.

Dry wells, interstitial sensors (on double-wall tanks or piping), and the outer piping or tank pitched toward a sump equipped with a leak sensor work well.

Exercise 5-6.

Electronic data loggers work well in this type of remote application because of their low power consumption, weather resistance, extensive data storage capacity, reliability, and relatively low cost.

Exercise 5-7.

Automatic sampling systems provide continuous, highly repeatable sampling, and greatly reduce the handling and risk of technician's exposure to any contaminants.

Exercise 5-8.

Level measurement devices for monitoring well applications include bubbler systems, head pressure transmitters, ultrasonics, capacitance or RF-type level detectors, and mechanical floats.

Unit 6

Exercise 6-1.

Environmental audits determine the nature and extent of contamination that has occurred within a predefined area.

Comprehensive audits may include a historical review of the area to determine whether the site has significant historical value and, therefore, should be left undisturbed. In addition, an investigation to determine whether the area is inhabited by an endangered species of animal or plant may be conducted.

Exercise 6-2.

Site audits may include air sampling, soil sampling, and groundwater analysis.

Exercise 6-3.

Preliminary soil borings and analysis are performed to determine the depth and area in which contamination exists. A sampling grid is established to determine overall area, and soil samples are obtained at varying depths at each sample point to determine the depth of contaminated soil at that location.

Exercise 6-4.

Weathered or nonvolatile soil samples make vapor analysis quite difficult. Sufficient vapor cannot be obtained from the samples to provide positive identification of the contaminants.

Exercise 6-5.

Typically, soil gas samples are obtained by the insertion of a perforated pipe into the soil and extraction of the gas occurs by pulling a vacuum on the perforated line. The discharged gases from the vacuum pump or blower are discharged into a collection vial or bag for analysis.

Removal of soil gases from the exhaust stream is accomplished by passing the gases through a carbon absorption bed or scrubber system. The remaining air is discharged into the atmosphere.

Exercise 6-6.

Thermal remediation involves heating the soil to a temperature that destroys the contaminant or fuses the soil into a solid block of harmless silicon or glass.

Disadvantage: Costly in terms of energy required. The soil is no longer in a granular form, which limits its use.

Another method uses a liquid poured into and absorbed by the soil, which solidifies into a solid block of material, sealing (encapsulating) the contaminant within so that it cannot leach into the groundwater.

Disadvantage: The contaminant remains present.

A third method involves destruction of the contaminants by microorganisms that feed on the contaminant and render it a harmless substance.

Disadvantage: Cost of process at this time. Cost driven by process required to expose soil to microorganisms and return soil to its original location.

Unit 7

Exercise 7-1.

Analysis of organics is frequently performed using a gas chromatograph in conjunction with infrared absorption or mass spectrometry.

For inorganics, inductively coupled plasma works well.

For slow dissolving solids, X-ray fluorescence is used.

For isomers, nuclear magnetic resonance is effective.

Exercise 7-2.

Electromagnetic focusing—The trajectory of the ions is detected through a slit in the shield ahead of the collector plate. The slit is positioned to allow the ions of the measured sample constituent to pass through based on its known trajectory.

Time-of flight—The ion gun is pulsed to allow the ions to be sorted based on their flight time between the gun and collector. The velocity of the ion varies with its mass according to the physics law:

$$\text{Energy} = 1/2mv^2$$

Radio frequency spectrometry—A series of grids is used to accelerate the ions and then alternate the polarity of the charge to aid or hinder the charge. The greater the ion mass, the more momentum and, therefore, the less effect the last grid has on its acceleration.

Exercise 7-3.

Isomers have the same mass; therefore, mass spectrometry cannot be used to distinguish between them.

Exercise 7-4.

The thermal conductivities (ability to conduct heat) of gases vary. Therefore, the thermal conductivity of an unknown gas can be compared to a reference gas with a known conductivity characteristic and identified.

Exercise 7-5.

When overlapping of peaks occurs, further analysis using a column with a different coating may further separate them. If a hit-and-miss approach using other columns does not separate them, it can be assumed that they are virtually identical substances and should be compared to a known substance with the same characteristics.

Exercise 7-6.

Most substances absorb differing levels of infrared radiation, allowing them to be identified. Refer to the OSHA chart in Appendix D.

INDEX